THE NEW PHYSICS
AND COSMOLOGY

THE NEW PHYSICS
AND COSMOLOGY

Dialogues with the Dalai Lama

Edited and narrated by Arthur Zajonc

with the assistance of Zara Houshmand

with contributions by David Finkelstein, George Greenstein, Piet Hut, Tu Weiming, Anton Zeilinger, B. Alan Wallace, and Thupten Jinpa

OXFORD
UNIVERSITY PRESS

2004

OXFORD

UNIVERSITY PRESS

Oxford   New York
Auckland   Bangkok   Buenos Aires   Cape Town   Chennai
Dar es Salaam   Delhi   Hong Kong   Istanbul   Karachi   Kolkata
Kuala Lumpur   Madrid   Melbourne   Mexico City   Mumbai   Nairobi
São Paulo   Shanghai   Taipei   Tokyo   Toronto

Copyright © 2004 by The Mind and Life Institute

Published by Oxford University Press, Inc.,
198 Madison Avenue, New York, New York 10016

www.oup.com

Oxford is a registered trademark of Oxford University Press

Library of Congress Cataloging-in-Publication Data
Bstan-'dzin-rgya-mtsho, Dalai Lama XIV, 1935 –
The new physics and cosmology : dialogues with the Dalai Lama /
edited and narrated by Arthur Zajonc with the assistance of
Zara Houshmand ; with contributions by David Finkelstein . . . [et al.].
p.   cm.
ISBN-13 978-0-19-515994-3
ISBN 0-19-515994-2
1. Physics—Religious aspects—Buddhism. 2. Quantum theory—Religious aspects—
Buddhism. 3. Buddhism—Doctrines. I. Zajonc, Arthur. II. Houshmand, Zara. III. Title.
BQ4570.P45B77 2003
294.3'375—dc21   2003048684

For further information concerning the Mind and Life Institute,
send an email to info@MindandLife.org or visit the following websites:
www.MindandLife.org
www.InvestigatingtheMind.org

9 8 7 6 5 4

Printed in the United States of America
on acid-free paper

ACKNOWLEDGMENTS

Over the years, Mind and Life Conferences have been supported by the generosity of many individuals and organizations.

Founders

Without the initial interest and continuous participation and support of our Honorary Chairman, His Holiness the Dalai Lama, the Mind and Life Institute would never have been formed, nor would it continue to flourish. It is truly extraordinary for a world religious leader and statesman to be so open to scientific findings and so willing to devote his time to creating and guiding a meaningful dialogue between science and Buddhism. Over the past fifteen years His Holiness has spent more personal time in Mind and Life dialogues than with any other non-Tibetan group in the world, and for this we are humbled, eternally grateful, and dedicate our work to his vision of seeing the richness of science and Buddhism linked in dialogue and scientific research collaboration, for the benefit of all beings.

Francisco J. Varela was our founding scientist, and we miss him enormously. Both a world-renowned scientist and a very serious practitioner of Buddhism, Francisco actually lived full time at the intersection of cognitive science and Buddhism, and was convinced that a deep and meaningful collaboration between science and Buddhism would be extremely beneficial for both systems, and for humanity itself. The direction he charted for the

Mind and Life Institute has been bold and imaginative, while at the same time respectful of the requirements of scientific rigor and Buddhist sensitivity. Above all, in this high-velocity world, he put aside time to cultivate the work of the Institute in a careful, logical, and scientifically incremental fashion. We continue on the road he set us upon.

R. Adam Engle is the entrepreneur who, upon hearing that His Holiness was interested in a dialogue between Buddhism and science, seized the opportunity and supplied the persistent effort and ingenuity to put the pieces in place for the work of the Institute to blossom and move forward.

Patrons

Barry and Connie Hershey of the Hershey Family Foundation have been our most loyal and steadfast patrons since 1990. Their generous support has not only guaranteed the continuity of the conferences, but it has also breathed life into the Mind and Life Institute itself.

Since 1990, Daniel Goleman has given generously of his time, energy, and spirit. He has prepared *Healing Emotions* and *Destructive Emotions* without compensation, as offerings to His Holiness the Dalai Lama and the Mind and Life Institute, who receive all the royalties from their publication.

We gratefully thank and acknowledge very generous multi-year support from Klaus Hebben, Tussi and John Kluge, Charlene Engelhard and the Charles Engelhard Foundation, Bennett and Fredericka Foster Shapiro, and the Sager Family Foundation. This critical, sustaining support enables the Mind and Life Institute to pursue its mission with continuity and vision.

The Institute has also received generous financial support from the Fetzer Institute, The Nathan Cummings Foundation, Branco Weiss, Stephen Friend, Marilyn and the late Don L. Gevirtz, Michele Grennon, Merck Laboratories, and Joe and Mary Ellyn Sensenbrenner.

The research projects of the Mind and Life Institute have received support from various individuals and foundations. Even though this support goes directly to the universities where the research is conducted, the Mind and Life Institute gratefully acknowledges and thanks the following donors for their generosity: The Fetzer Institute, John W. and Tussi Kluge, Charlene Engelhard and the Charles Engelhard Foundation (UCSF Medical Center), and Edwin and Adrianne Joseph (University of Wisconsin).

Finally, we gratefully thank the Sager Family Foundation, which has generously supported the science education of Tibetan monks in India on a multi-year basis.

On behalf of His Holiness the Dalai Lama, and all the other participants over the years, we humbly thank all of these individuals and organizations. Their generosity has had a profound impact on the lives of many people.

Scientists and Philosophers

We would also like to thank a number of people for their assistance in making the work of the Institute itself a success. Many of these people have assisted the Institute since its inception. First and foremost we thank His Holiness the Dalai Lama and the scientists, philosophers, and Buddhist scholars who have participated in our past meetings, our current meetings, our research projects, our board of directors, and our scientific advisory board: the late Francisco Varela, Richard Davidson, Daniel Goleman, Anne Harrington, Jon Kabat-Zinn, Thupten Jinpa, Bennett Shapiro, Alan Wallace, Arthur Zajonc, Paul Ekman, Pier Luigi Luisi, Matthieu Ricard, Evan Thompson, the late Robert Livingston, Newcomb Greenleaf, Jeremy Hayward, Eleanor Rosch, Patricia Churchland, Antonio Damasio, Allan Hobson, Lewis Judd, Larry Squire, Daniel Brown, Clifford Saron, Sharon Salzberg, Lee Yearley, Jerome Engel, Jayne Gackenbach, Joyce McDougall, Charles Taylor, Joan Halifax, Nancy Eisenberg, Robert Frank, Elliott Sober, Ervin Staub, David Finkelstein, George Greenstein, Piet Hut, Tu Weiming, Anton Zeilinger, Owen Flanagan, Mark Greenberg, Jeanne Tsai, Ajahn Maha Somchai Kusalacitto, Michael Merzenich, Steven Chu, Ursula Goodenough, Eric Lander, Michel Bitbol, Phillip Sharp, Jonathan Cohen, John Duncan, David Meyer, Anne Treisman, Ajahn Amaro, Daniel Gilbert, Daniel Kahneman, Dacher Keltner, Georges Dreyfus, Stephen Kosslyn, Marlene Behrmann, Daniel Reisberg, Elaine Scarry, Jerome Kagan, Antoine Lutz, Gregory Simpson, Margaret Kemeny, Sogyal Rinpoche, Tsoknyi Rinpoche, Mingyur Rinpoche, and Rabjam Rinpoche.

The Private Office and Tibetan Supporters

We thank and acknowledge Tenzin Geyche Tethong, Tenzin N. Taklha, Ven. Lhakdor, and the other wonderful people of the Private Office of His Holiness. We are grateful to Rinchen Dharlo, Dawa Tsering, and Nawang Rapgyal of the Office of Tibet in New York City, and Lodi Gyari Rinpoche of the International Campaign for Tibet for their help over the years. And special thanks to Tenzin Choegyal, Ngari Rinpoche, who is a board member, a wonderful guide, and a true friend.

Other Supporters

Our thanks to Kashmir Cottage, Chonor House, Pema Thang Guesthouse and Glenmoor Cottage in India, Maazda Travel in the United States and Middle Path Travel in India, Elaine Jackson, Zara Houshmand, Alan Kelly, Peter Jepson, Pat Rockland, Thupten Chodron, Laurel Chiten, Billie Jo Joy, Nancy Mayer, Patricia Rockwell, George Rosenfeld, Andy Neddermeyer, Kristen Glover, Maclen Marvit, David Marvit, Wendy Miller, Sandra Berman, Will Shattuck, Franz Reichle, Marcel Hoehn, Geshe Sopa and the monks and nuns of Deer Park Buddhist Center, Dwight Kiyono, Eric Janish, Brenden Clarke, Jaclyn Wensink, Josh Dobson, Matt McNeil, Penny and Zorba Paster, Jeffrey Davis, Magnetic Image, Sincerely Yours, Health-Emotions Research Institute-University of Wisconsin; Harvard University's Mind/Brain/Behavior Interfaculty Initiative, Karen Barkow, John Dowling, Catherine Whalen, Sara Roscoe, David Mayer, Jennifer Shephard, Sydney Prince, Metta McGarvey, Ken Kaiser, Gus Cervini, Marie Seamon, T&C Film, Shambhala Publications, Wisdom Publications, Oxford University Press, Bantam Books, and Snow Lion Publications.

Interpreters

Finally, our very special thanks go to our interpreters over the years: Geshe Thupten Jinpa, who has interpreted for every meeting; Alan B. Wallace, who has been with us for every meeting but one; and Jose Cabezon, who pitched in for Alan while he was on retreat in 1995. As you can imagine, creating a dialogue and collaboration between Tibetan Buddhists and western scientists is a nonstarter without excellent translation and interpretation. These friends are, quite literally, the best in the world.

CONTENTS

THE PARTICIPANTS

TENZIN GYATSO, HIS HOLINESS, THE FOURTEENTH DALAI LAMA, is the leader of Tibetan Buddhism, the head of the Tibetan government in exile, and a spiritual leader revered worldwide. He was born to a peasant family on July 6, 1935, in the small village of Taktser in northeastern Tibet. He was recognized at the age of two, in accordance with Tibetan tradition, as the reincarnation of his predecessor, the Thirteenth Dalai Lama. The Dalai Lamas are believed to be the manifestations of the Buddha of Compassion, who chooses to reincarnate for the purpose of serving humankind. Winner of the Nobel Prize for Peace in 1989, he is universally respected as a spokesman for the compassionate and peaceful resolution of human conflict. He has traveled extensively, speaking on such subjects as universal responsibility, love, compassion, and kindness. Less well known is his intense personal interest in the sciences and their implications; he has said that if he were not a monk, he would have liked to be an engineer. As a youth in Lhasa, it was he who was called on to fix broken machinery in the Potala Palace, be it a clock or a car. He has a vigorous interest in learning about the newest developments in science and brings to bear both a voice for humanistic implication of the findings and a high degree of intuitive methodological sophistication.

DAVID RITZ FINKELSTEIN teaches and studies physics at the Georgia Institute of Technology and edits the *International Journal of Theoretical Physics*. When he learned in college that quantum physics revises the logic for physical systems, he began working to extend quantum logic to still deeper levels of physics. As byproducts of this main interest, he has contributed to early work on the topology of the gravitational field, the concept of the black hole, the gauge theory of the electroweak interactions, and quantum theory. He has elaborated the Copenhagen epistemology into a relativistic philosophy, which he calls *practic*, based on processes rather than states. He currently explores the consequences of a process atomic hypothesis: that all physical processes are made up of finitely many indivisible elementary ones; see his book, *Quantum Relativity* (New York: Springer, 1996).

GEORGE GREENSTEIN is the Sidney Dillon Professor of Astronomy at Amherst College. He received his B.S. from Stanford University and his Ph. D. from Yale University, both in physics. Initially his interests centered on research in theoretical astrophysics but later shifted to writing. He is the author of numerous works, interpreting science for nonscientists. His first book, *Frozen Star* (New York: Freundlich, 1983), was the recipient of two science-writing awards. He is also author of *The Symbiotic Universe: Life and Mind in the Cosmos* (New York: Morrow, 1988) and *Portraits of Discovery: Profiles in Scientific Genius* (New York: Wiley, 1998). In conjunction with Arthur Zajonc, he is the author of a textbook entitled *The Quantum Challenge: Modern Research on the Foundations of Quantum Mechanics* (Sudbury, MA: Jones and Bartlett, 1997), which discusses the problems of interpretation posed by quantum mechanics.

PIET HUT is professor of astrophysics and interdisciplinary studies at the Institute for Advanced Study in Princeton, where he has been since 1985. He received his Ph.D. at the University of Amsterdam, Holland. For many years he has been involved in a Tokyo-based project aimed at developing a special-purpose computer for simulations in stellar dynamics, with a speed of one Petaflops. He is famous for inventing the Barnes-Hut algorithm for efficient calculation of interparticle distances and dynamics for the many-body problem. Besides his work in theoretical astrophysics, much of his research has a broadly interdisciplinary character: He has coauthored articles with computer scientists, particle physicists, geologists, paleontologists, psychologists, and philosophers. During the last few years, he has organized a series of workshops to investigate the character of intrinsic limits to scientific knowledge. He has focused on three main questions: To what extent can limits be seen as dictated by the structure of human knowledge? To what extent are limits given in the structure of nature itself? And to what extent are limits inherent in any attempt to map reality into a model? Since 1996, Piet

Hut has been president of the Kira Institute, which explores the relationship among science, ethics, and aesthetics from a nonreductive viewpoint.

THUPTEN JINPA was born in Tibet in 1958. Trained as a monk in southern India, he received the *geshe lharam* degree (equivalent to a doctorate in divinity) from Shartse College of Ganden Monastic University, where he also taught Buddhist philosophy for five years. He also holds a B.A. (honors) in Western philosophy and a Ph.D. in religious studies, both from Cambridge University. Since 1985 he has been a principal English translator to His Holiness the Dalai Lama and has translated and edited several books by the Dalai Lama, including *The Good Heart: A Buddhist Perspective on the Teachings of Jesus* (Boston: Wisdom, 1996) and *Ethics for the New Millennium* (New York: Riverhead, 1999). His most recent works are (with Jaś Elsner) *Songs of Spiritual Experience* (Boston: Shambala, 2000), the entries on Tibetan philosophy in the *Encyclopedia of Asian Philosophy* (New York: Routledge, 2001), and *Self, Reality and Reason in Tibetan Philosophy: Tsongkhapa's Quest for the Middle Way* (New York: Routledge/Curzon, 2002). From 1996 to 1999 he was the Margaret Smith Research Fellow in Eastern Religion at Girton College, Cambridge University. He is currently the president of the Institute of Tibetan Classics, which is dedicated to translating key Tibetan classics into contemporary languages. He lives in Montreal, Canada, with his wife and two young children.

B. ALAN WALLACE trained for many years in Buddhist monasteries in India and Switzerland, and he has taught Buddhist theory and practice in Europe and America since 1976. He has served as interpreter for numerous Tibetan scholars and contemplatives, including His Holiness the Dalai Lama. After graduating summa cum laude from Amherst College, where he studied physics and the philosophy of science, he earned a doctorate in religious studies at Stanford University. He has been a visiting professor at the University of California in Santa Barbara. He has edited, translated, written, or contributed to more than thirty books on Tibetan Buddhism, medicine, language, and culture, as well as the relationship between science and religion. His published works include *Tibetan Buddhism from the Ground Up* (Boston: Wisdom Publications, 1993), *Choosing Reality: A Buddhist View of Physics and the Mind* (Ithaca, N.Y.: Snow Lion Publications, 1996), *The Bridge of Quiescence: Experiencing Buddhist Meditation* (Chicago: Open Court, 1998), and *The Taboo of Subjectivity: Toward a New Science of Consciousness* (New York: Oxford University, 2000). His forthcoming anthology of articles is entitled *Buddhism and Science: Breaking New Ground* (New York: Columbia University, 2003).

TU WEIMING, director of the Harvard-Yenching Institute, was born in February 1940 in Kunming, China. He earned his B.A. degree in Chinese studies from Tunghai University, Taiwan. He received his M.A. in religious studies from Harvard University in 1963 and his Ph.D. in history and East Asian languages in 1968, also from Harvard. He has taught at Princeton University and the University of California at Berkeley, and since 1981 he has held the position of professor of Chinese history and philosophy at Harvard University. Active in many public bodies, he is a member of the Committee on the Study of Religion at Harvard, the chair of the Academia Sinica's advisory committee on the Institute of Chinese Literature and Philosophy, and a fellow of the World Economic Forum regularly held in Davos, Switzerland. He is a member of the Group of Eminent Persons on the Dialogue among Civilizations convened by the secretary general of the United Nations, a fellow of the American Academy of Arts and Sciences, and a board member of the Chinese Heritage Centre in Singapore. In 1999 he was awarded the title of Harvard-Yenching Professor of Chinese History and Philosophy and of Confucian Studies. He was awarded an honorary doctorate from Lehigh University in 2000 and in 2001 received the Ninth International T'oegye Studies Award from the T'oegye Studies Institute in Seoul, South Korea. He is the author or editor of 19 books in English, 13 books in Chinese, and well over 100 articles and book chapters.

ARTHUR ZAJONC is professor of physics at Amherst College, where he has taught since 1978. He received his B.S. and Ph.D. in physics from the University of Michigan. He has been visiting professor and research scientist at the École Normale Supérieure in Paris, the Max Planck Institute for Quantum Optics, and the universities of Rochester and Hanover. He has been a Fulbright professor at the University of Innsbruck in Austria. As a postdoctoral fellow at the Joint Institute for Laboratory Astrophysics, he researched electron-atoms collision physics and radiative transfer in dense vapors. His research has included studies in parity violation in atoms, the experimental foundations of quantum physics, and the relationship between sciences and the humanities. He has written extensively on Goethe's science. He is the author of *Catching the Light: The Entwined History of Light and Mind* (New York: Oxford University, 1995), coauthor of *The Quantum Challenge: Modern Research on the Foundations of Quantum Mechanics* (Sudbury, MA: Jones and Bartlett, 1997), and coeditor of *Goethe's Way of Science: A Phenomenology of Nature* (Albany: State University of New York, 1998).

He is a founding member of the Kira Institute, which explores the relationships among science, values, and spirituality. He is a consultant with the Fetzer Institute and president of the Anthroposophical Society in America and the Lindisfarne Association.

ANTON ZEILINGER completed all his studies at the University of Vienna. He was director of the Institute for Experimental Physics and professor of physics at the University of Innsbruck in Austria from 1990–99. He is currently professor of experimental physics at the University of Vienna. He has been a visiting professor at the Collège de France in Paris and at Merton College, Oxford University. He served as president of the Austrian Physical Society from 1996–98 and is a member of the Austrian Academy of Sciences. He has been awarded numerous prizes for his work in physics, including the European Optics Prize in 1996, the Senior Humboldt Fellow Prize in 2000, and the Science Prize of the City of Vienna in 2000. His research interests are in the foundations of quantum physics. His research group aims to demonstrate the novel counterintuitive quantum phenomena through experiment. This work is paralleled by theoretical investigations into the structure of quantum mechanics and epistemological investigations into the kinds of statements about the world that can be made in view of quantum physics. The recent achievement of quantum teleportation attracted worldwide attention.

THE NEW PHYSICS
AND COSMOLOGY

PRELUDE

The morning after our arrival in Dharamsala, India, home to many Tibetans in exile, I made my way by foot along narrow, rutted roads until I came to the Tibetan Children's Village. Nestled in the foothills of the Himalayas, the orphanage and schools of the Children's Village make up a small world of some 2,500 refugee children, teachers, and caregivers striving to preserve their ancient culture while simultaneously becoming part of modern civilization. Nearby is the residence and monastery of His Holiness the Fourteenth Dalai Lama. During the coming week, from October 27 to 31, 1997, five other scientists and I would be conversing with him about our intersecting interests, Buddhist philosophy and modern physics.

At the Children's Village, in an open-air pavilion that did little to shield us from the cold October mists, a classic Tibetan opera was underway. Magnificently costumed singers and dancers performed in a style that seemed a strange combination of ancient fairy tale and classical Asian theater, with a dash of slapstick that invariably brought wide smiles and laughter to the crowd. Rather abruptly, a pause in the performance was announced and whispers went through the audience. More people emerged from nearby buildings to join the throng. With only a few monks, trying helplessly to protect him from the drizzle, the Dalai Lama made his way down the long stairway, bowing to all around him and grasping outstretched hands in both of his, his face bright with the infectious smile known around the world. For the next week he would be discussing quantum physics and cosmology with us, but this morning he was here at the

3

Children's Village, offering words of encouragement and blessing to each and all.

Five days later, following the close of our meetings with the Dalai Lama, we were all gathered—scientists, philosophers, monks, and friends—on the balcony of the small monastery that is part of the Dalai Lama's compound. Again the sky was gray and rainy. As we talked, the heavens brightened and a full, glorious rainbow arched between the mountains and us.

These two events—a Tibetan opera and a rainbow—were the bookends to a remarkable set of conversations that my colleagues and I were privileged to have with the Dalai Lama about the new physics and cosmology. This book is the record of those conversations.

The new physics and cosmology of the twentieth century are replete with understandings of our universe that challenge nearly every classical scientific notion we have inherited from the nineteenth century. Scientific titans, such as Galileo and Newton, Copernicus and Kepler, Faraday and Maxwell, fashioned that viewpoint. Their method of inquiry, as well as their understanding of the universe, was profoundly different from that practiced by medieval and ancient natural philosophers. The new science was predicated on experiment, systematic observation, and theoretical models of a novel type. The success of their style of science, as gauged both by its predictive power and its technical applications, was astounding. Newton's theory of dynamics was applied to the intricate phenomena of the heavens and explained the motions of the planets and stars according to the same laws that governed terrestrial motion, something thought to be impossible by the ancient Greek philosophers. Optics was joined to the new science of electromagnetism, providing a profound field-theoretic view of electrical and magnetic forces and, by analogy, even gravity. The success of physical science was such that by the end of the nineteenth century Lord Kelvin, among others, announced that the universe in its entirety had been fathomed, and only the uninteresting details remained. He had enough wit to recognize two "clouds" on the horizon that did not fit into his optimistic scenario: the failure of Michelson and Morley's search for the ether and the failure of theory to predict the spectrum of light given off by matter at high temperatures. The first cloud gave rise to relativity and the second to quantum mechanics. Lord Kelvin was prescient, if also arrogant.

During the three centuries that established classical physics and cosmology, the mechanistic and materialistic character of physical theory came to dominate Western thinking even outside these areas. Increasingly, philosophy came under the powerful sway of science through such thinkers as Descartes, Kant, and Locke. The life sciences, longing for comparable precision, sought out a similar path of development to that of physics. Genetics, evolution, and cellular biology displaced natural history and whole-organism biology. The mind itself, traditionally understood as the expression of the spirit, gradually became part of the mechanistic universe as well. By the dawn of the twentieth century, the physics of the seventeenth century had successfully conquered the adjacent areas of science and was already encroaching on that of the mind. A single mechanistic paradigm and its associated materialistic metaphysics came to dominate Western thinking.

With the opening of the twentieth century, the theories of quantum mechanics and relativity would make incomparable demands on our conception of the universe. We are still struggling to grasp their full implications. They challenge the simple mechanistic accounts of matter and the cosmos we inherited from earlier centuries, replacing them with accounts that shun such pictures. In addition, both quantum theory and relativity grant a new prominence to the observer. It is hard to overestimate the significance of these developments. The ramifications of twentieth-century discoveries for physics and cosmology have been enormous, changing our very notions of space and time, the ultimate nature of matter, and the evolution of the universe. They have also begun to affect philosophical discussions in significant ways.

While the philosophical implications of the new physics are still being sorted out in the West, what better topic to discuss with Buddhism's leading representative? As the spiritual leader of the Tibetan people, the Dalai Lama is well schooled in the intricacies of Tibetan Buddhist philosophy, epistemology, and metaphysics. We were all anxious to present to him the conceptual revolution instigated by modern physics and to analyze with him its philosophical implications. Although Buddhism has little experience with the specific theories of modern science, it has long inquired into the fundamental nature of substance and the nature of the mind; it has thought deeply about experience, inference, causality, and the proper role of concepts and theories in our thinking. Even the long history of the physical universe has been the subject of Buddhist reflection, leading to remarkable views not unlike those being advanced today by cosmologists.

In these dialogues, the reader has the rare opportunity of learning about the new physics and cosmology together with one of Asia's deepest philosophical thinkers. We quickly discovered that although the Dalai Lama lacked formal instruction in physics, he was a brilliant student, often anticipating our next remarks and posing penetrating questions. Each morning, under a continual stream of inquiries from the Dalai Lama, one of the scientists—three physicists and two astrophysicists—tutored him in the discoveries science has made in the areas of quantum mechanics, relativity, and modern cosmology. Each afternoon, our conversations were of a freer nature, drawing their subject matter from the striking philosophical implications of the morning's topic. In these exchanges we were much helped by the contributions of Harvard philosopher and Asian historian Tu Weiming, whose understanding of Eastern, as well as Western, philosophy provided broad and illuminating viewpoints.

Time and again throughout our five days together, the dialogue would grow intense as we all attempted to understand more fully the paradoxical features of the new physics and cosmology. The Dalai Lama was a full participant in our conversations. Indeed, by the end of our time together, Austrian physicist Anton Zeilinger went so far as to speak appreciatively of the Dalai Lama as a genuine scientific collaborator and to invite him to his Innsbruck laboratory. During June 1998, Anton and I enjoyed a three-day visit from His Holiness in Innsbruck, where Anton was able to show him the actual experiments that support the startling conclusions of quantum theory and where we continued our probing conversations about the foundations of quantum mechanics. The Innsbruck conversations, however, will have to wait for another book.

The Dalai Lama is not only the secular leader of Tibet in exile but also the leader of Tibetan Buddhism. One might justifiably ask, On what sound intellectual basis can scientists have a dialogue with religious leaders? After all, religions are characterized by faith in particular doctrines, whereas science attempts to discover laws of nature by means of careful observation, experimentation, and reason. In the Dalai Lama's opening remarks to us, however, it became clear that a deep commitment to careful inquiry and valid cognition are also at the heart of Buddhist philosophy.

DALAI LAMA: In Buddhism in general, and particularly in Mahayana Buddhism, the basic attitude is that you should remain skeptical at the beginning. Even the Buddha's own words say that it is better to remain skeptical. This skeptical attitude automatically brings up questions. Questions bring clearer answers, or investigation. Therefore, Mahayana Buddhist thinking relies more on investigation rather than on faith. I feel that that attitude is very, very helpful in communicating with scientists.

Buddhist ethical discourse often speaks about wrong views as constituting a negative state of mind. There are two kinds of wrong views: One exaggerates what is actually there, superimposing onto a thing a property of existence or status that is not there. The other denies what is actually there. So both absolutism and nihilism are seen as wrong views. Thus even in ethical discourse, a correct understanding of reality is very much emphasized. Therefore, scientific findings are very helpful to Buddhist thinking.

Some Buddhist views also give scientists a new way of looking, as I've found in my past experience. Some scientists have an interest or enthusiasm to learn more about Buddhist explanations in their particular field. Because of this, I feel that my meetings with scientists are very useful and productive. Given that science as a discipline and Buddhism as a system of thought both share a basic commitment to openness and initial skepticism, it is important that all the participants have an understanding that there should be total openness in our discussions, and a free exchange of ideas with no preset rules.

With these remarks to guide our conversation, we could begin in real earnest. No subject was off-limits. Hard questions could be asked from both sides. For all the differences between Western science and Buddhist philosophy, the Dalai Lama repeatedly demonstrated his commitment to careful analytical reasoning and to the crucial role of experience. We were all committed to the same goal: finding the truth. For Buddhism, ignorance is understood as the root cause of suffering because a mistaken view of the world or of the self inevitably leads to attachments and destructive emotions. Truth is thus essential to a Buddhist's goal: the reduction of suffering. The sciences also seek truth, not only as an end in itself but also to alleviate illness and suffering through the ethical application of technology. By bringing the greatest accomplishments of Western science together with the most skillful thinking and philosophical insights from Tibet, we hoped to shed some light on the thorny issues of modern physics that have so far eluded our understanding. We did not expect final solutions but rather sought fresh approaches to old problems. Early in our discussion, Tu Weiming spoke directly to the hopes of those present:

> Many of the great accomplishments in modern Western science became highly problematic because of the new developments in physics. We are at a stage where new knowledge will have to come from a much broader collaborative effort. That collaborative effort may involve people from many different disciplines and different traditions but with a precision that has been advanced by science.

Around the table in Dharamsala was seated a variety of disciplines and traditions, just as Tu Weiming had imagined.

Anton Zeilinger was there from the University of Innsbruck, where he led a renowned experimental group that probed the foundations of quantum mechanics. While a Fulbright professor at Innsbruck, I appreciated the unique blend of cutting-edge experiments and the subtle philosophical discussions that characterized his research group. Winner of numerous international awards for his physics research, Anton's work spans three related areas in the foundations of quantum physics: the interference of neutrons, the interference of atoms (including the molecule C_{60}), and the study of photons. His group was the first to teleport the quantum state of a photon, developing the theory and experiments for new tests of quantum nonlocality; its members have been active in the emerging field of quantum information processing with its promise of quantum computers and quantum cryptography. Now at the University of Vienna, Anton continues his research as professor of experimental physics. On the first day, Anton opened our session with an introduction to the primary questions posed by quantum experiments.

David Finkelstein, from the Georgia Institute of Technology, added to the proceedings his remarkable mastery of relativity, quantum theory, and quantum logic. Editor of the *International Journal of Theoretical Physics* for twenty-five years, author of many important theory papers and the book *Quantum Relativity: Synthesis of the Ideas of Einstein and Heisenberg*, David brought to the table a widely respected theoretical mind. His sense of irony and precision was appreciated, especially because his area of presentation on the second day was the most difficult of our week.

As the scientific organizer of the meeting, I had the twin responsibilities of presenting and facilitating the dialogue. My own background was in experimental atomic and optical physics, at first as a postdoctoral fellow at the Joint Institute for Laboratory Astrophysics and then at Amherst College. Since 1980 I had become increasingly interested in the role of experiment in demonstrating the conceptual puzzles of quantum mechanics. In the early 1980s this field had involved only a handful of experimentalists, but since then it has grown enormously, with many groups performing experiments all over the world. I had studied the subtleties of measurement through the so-called quantum eraser while at the École Normale Supérieure. I also collaborated on an experiment at the Max Planck Institute for Quantum Optics in Munich that implemented John Archibald Wheeler's famous delayed-choice experiment. Parallel with my work in physics, I had consistently pursued a second line of research into the historical and philosophical dimensions of physics, including the relationship of science to our ethical and spiritual concerns. This culminated in my book *Catching the Light: The Entwined History of Light and Mind*. Although not a Buddhist myself, I had come to appreciate the care and depth of its philosophical system and contemplatively based "inner science," and so I looked forward to the opportunity of discussing physics in the broader context of Buddhist philosophy.

On the last two days, astrophysicists George Greenstein from Amherst College and Piet Hut from the Princeton Institute for Advanced Study would lead us into the latest thinking and ongoing debates within cosmology. George and I have been colleagues for many years at Amherst, where he is a highly respected teacher, writer, and researcher. Since graduating from Yale and Stanford Universities, George has focused his research on neutron stars, pulsars, and the big bang, but his real love is seminar-style teaching, and he is a leader in the astrophysical community in this area. His book *Frozen Star,* on black holes, neutron stars, and other exotic astronomical objects, won major awards for its science writing. Explaining the universe to the nonspecialist is George's specialty, and we would need it if we were to get across to the Dalai Lama the ideas of curved space-time in general relativity and the early inflation of the universe.

Piet Hut holds the unique distinction of being a professor of both astrophysics and interdisciplinary studies at the Institute for Advanced Studies in Princeton, one of America's most prestigious research institutions. Piet distinguished himself early for his landmark work on cosmological neutrinos, as well as for modeling the dynamics of the millions of stars that make up globular clusters. He and colleagues designed and used the world's fastest special-purpose computer to do their modeling of colliding galaxies. In the last several years, Piet has increasingly extended his research and writing to include philosophy, being influenced especially by the phenomenological approach of Edmund Husserl. On the final day of our time together, Piet would bring both aspects of his work to the table: cosmology and philosophy. After discussing the evolution of the stars, he sought a way to bring the values dimension of experience into our scientific account of reality. In this way we squarely confronted the complex relationship between religion and science.

Tu Weiming was born in Kunming, China, and studied in Taiwan. He is professor of Chinese history and philosophy at Harvard and director of the Yenching Institute. The Dalai Lama had long wished for representation from China in the dialogues, as he was always searching for ways to cross the barriers generated by the invasion of Tibet. Weiming played an essential role by helping us to bridge the difference in the intellectual and spiritual cultures of Asia and the West.

Finally, I must include in my remarks on the participants a few words about our two interpreters, Thupten Jinpa and B. Alan Wallace. Although the Dalai Lama's English is quite good, when tackling difficult scientific and philosophical material he often asked for translation both into and out of Tibetan. But Jinpa and Alan were also fully trained in Tibetan Buddhism (both having been monks for many years) and well schooled in Western philosophy. Alan Wallace had studied physics with me at Amherst College, after which he completed his Ph.D. in religious studies at Stanford. Thupten Jinpa had completed his geshe degree (equivalent to our Ph.D. in theology) before coming to Cambridge University, where he completed a B.A. (with honors in philosophy) and a Ph.D. in religious studies. In addition to translating, Alan and Jinpa often acted as consultants to the Dalai Lama as he developed his own responses to the scientific material. One must really consider these two scholars as full participants, making our group a circle of nine.

Evident throughout our conversations was a genuine respect for the viewpoints of each individual, which led in turn to a wonderful mood of collaborative inquiry. Around the table sat representatives from all domains of twentieth-century physical science and Tibetan Buddhism, as well as the Dalai Lama. Everything was ready for a wonderful conversation. All that remained was to begin. No book can do justice to the lively human dimensions of the meeting, but perhaps between the lines one can sense the passion and puzzlement, the humor and hospitality, that occurred throughout our time together. It was not completely unlike the opera I witnessed on arriving. The scene was at once ancient and modern, with monks in traditional dress and laboratory equipment on the table before us. Laughter and earnest, energetic debate alternately filled the room. In place of the audience of children and villagers, fifty invited guests attended, each an accomplished student of either philosophy or science. I can't promise a rainbow for the ending, but maybe the reader will be able to weave one out of the many-hued strands of our wide-ranging considerations.

I

Experiment and Paradox
in Quantum Physics

On the first day of our meetings, the Austrian experimental physicist Anton Zeilinger was asked to introduce the Dalai Lama to the fundamental features of quantum mechanics. Anton is one of the world's foremost experts in the field of experimental foundations of quantum mechanics. He is probably best known for his groundbreaking experiments that demonstrate quantum teleportation, or the transmission of an exact replica of an arbitrary quantum state to a distant location. For his session, Anton had brought to India a complex and highly miniaturized quantum experiment with which he was able to demonstrate the central mysteries of quantum mechanics.

In short order, Anton introduced the Dalai Lama to wave-particle duality for single photons, to the concept of objective randomness in quantum mechanics, and to the profound mystery of nonlocality for two-particle systems. In every instance, Anton attempted to stay as close to the phenomena of quantum experiments as possible, using the minimum number of presuppositions in his arguments. This was central to his philosophical viewpoint. Not surprisingly, therefore, a prominent theme in our conversations turned on the role of the observer in experiments and the dangers of using models to picture the workings of quantum systems.

At that time, Anton worked at the University of Innsbruck, which, like Dharamsala, is situated in a magnificent mountain landscape. As we gathered around the long coffee table, Anton began his morning presentation with an appreciation of the openness of the Dalai Lama to new knowledge and with a slide picture of the Tyrolian Alps.

ANTON ZEILINGER: Your Holiness, the skepticism that you remarked on is exactly what drives us in science. Only if you are skeptical of what somebody tells you—no matter how famous or important he is—only then can you learn something new. It is the only road to new knowledge.

I show this picture of a mountain to remind us of our view of everyday life, including the view from classical physics. In everyday life we usually don't doubt whether the mountains are there when we aren't looking at them. One can question these things from a philosophical viewpoint, but in classical physics and in everyday life the mountain is there even when I don't look. In quantum physics, this position no longer works. In the next hour or so, I want to give you some of the reasons why we believe this. I'll do this by discussing the nature of light because light was a driving force in the development of these ideas.

A very important observation was made in the year 1802 by an English medical doctor, Thomas Young, who did the famous double-slit experiment. I brought with me a version of the experiment with modern-day equipment [see figure 1.1]. A little laser here emits red light. There is a barrier here with two slits open, side by side. The light com-

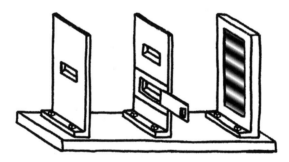

Figure 1.1 Double-slit experiment. Light is incident from the left and passes through the first slit. It then encounters a screen with two slit openings, one of which can be closed by a small shutter. Finally, the light falls on an observation screen. When both slits are open, we observe bright and dark stripes. When one slit is closed, no stripes occur and we observe a homogeneous medium brightness on the observation screen. The stripes are due to superposition of the light waves passing through the two slits. The conceptual question discussed in the text arises when one realizes that light is composed of individual quanta of light, called photons, that display particle properties.

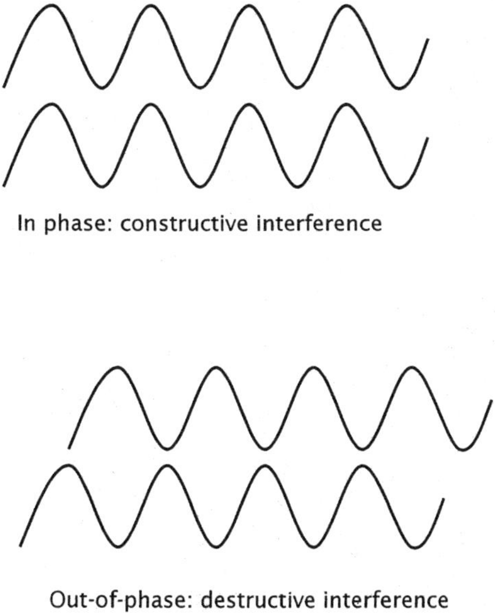

In phase: constructive interference

Out-of-phase: destructive interference

Figure 1.2 Waves that are in phase constructively interfere; those that are out of phase destructively interfere.

ing through the slits throws a pattern on the screen in back. The important point here is the distribution of the light. You see bands of red light, with dark stripes between them. If I close one slit, you see that the black stripes disappear. With one slit closed, the light pattern is homogeneous. Now, with both slits open, suddenly the black stripes appear.

This was a very important observation in the history of physics. Why is it important? How can we understand this? There was a discussion for a long time in the history of physics as to whether light is a wave of some kind or whether it is particles, little pieces of something. This experiment seemed to demonstrate that light is a wave.

When two waves meet, their oscillations interact in some way. Two extremes can be observed [*see figure 1.2*]. At one extreme, they oscillate out of phase, contrary to each other. When the two waves meet, they extinguish each other. You can see this in water waves. At the other extreme, if they happen to be oscillating the same way, the two waves then reinforce each other. The two slits allow two possible paths for the light. The waves going through each slit have to travel slightly different distances to reach the same point on the screen. The different path lengths and different travel times mean that the waves from each slit will oscillate the same way at certain points on the screen, producing the light bands, and at other points will extinguish each other, producing the dark bands.

DALAI LAMA: Would this work equally well if that laser were a different color than red? If it were blue or yellow?

ANTON ZEILINGER: Yes. We now know that all waves work this way—light, water waves, radio waves, even atoms. Atoms can also have a wave nature. People have done the same experiment by sending atoms through a two-slit assembly, leading to the same phenomenon. It is a universal phenomenon.

DALAI LAMA: I'm picturing waves of light moving like water waves, in constant motion. If you could look with microscopic precision, would you see the movement reflected there on the surface? Would the black and red lines appear to fluctuate or vibrate? Or would they be completely static?

ANTON ZEILINGER: Your eye is very, very slow. These waves are oscillating at a frequency of 100 million multiplied by 1 million times per second. It is too fast to see.

DALAI LAMA: (laughing) I half believe you. . . . I'm taking a skeptical view!

Wave-Particle Duality

ANTON ZEILINGER: What physics learned from Thomas Young's experiment is that we can understand light as a wave. That was the complete view until a new experiment was done at the beginning of our century. People observed that when they shined light on a metal plate, under certain circumstances, electrons—small elementary particles—are emitted from the metal.

The phenomenon was known for some time as an experimental observation, but it was not understood until Albert Einstein explained it in 1905. (It was this that won Einstein the Nobel Prize. His relativity theory was considered too radical for the Nobel Prize, but this is actually no less radical.) Einstein said there is a very simple way to understand it: Let us assume that light is made of particles, which later came to be called photons. When light reaches a metal surface, sometimes a photon kicks out an electron, just as one ball kicks out another ball. Not only was it a very simple picture, but it also could explain certain quantitative predictions, like how fast the electrons moved.

I have an experiment that demonstrates this for you. Here we have a little box, which contains a photon detector. It is a metal plate which the light can hit, and we can then electrically register the electrons re-leased. A little loudspeaker here makes a click every time a photon is de-tected. I hope it works; you never know.

At this point Anton opened a shutter covering the light-sensitive surface of the photon detector. When opened, a clicking sound was heard; when closed, there was silence. The Dalai Lama suggested that Anton take the box into the morning sunlight that was streaming through the window. When he did so, the clicking became more rapid. The clicking supported Einstein's 1905 hypothesis that photons, when detected, act like particles, kicking out electrons from the metal surface.

ANTON ZEILINGER: So, there was an interesting situation in physics. We had two pictures of light: the wave picture and the particle picture. For a long time the question was how to understand the two.

DALAI LAMA: Do the photon particles actually displace the electrons like two billiard balls? Or, since they are different types of particles, can the displacement occur without actual physical contact?

ANTON ZEILINGER: The question is a very hard one. The reason is that in quantum physics, we have given up such pictures. We can describe the phenomenon, up to a certain extent, *as if* this particle kicks out the other particle. But we have learned now that we really should only talk about the phenomena we can observe.

DALAI LAMA: There is a problem here. You say that we have two pictures of light, particle and wave, but when I ask you this question, then you say we have no picture.

ANTON ZEILINGER: We have two pictures which are conflicting. We know today (we did not know this in Einstein's time) that both pictures should only be used to help us see a little bit of what's going on. But both are really not adequate. We should not have pictures anymore.

DALAI LAMA: Can you explain why a single phenomenon cannot be both a particle and a wave? What are the mutually exclusive properties of particles and waves?

ANTON ZEILINGER: I would like to explain this with a demonstration that underlines the problem we have here. If it still works; it has a bad con-tact. This is what we have to work with experimentally. We spend most of the time fixing problems like this.

Anton repaired the photon detector and then placed it behind the barrier with the two slits. The clicking registered the light coming through the slits.

ANTON ZEILINGER: So . . . the light coming through the two slits is also made of particles. But how can we understand what is going on? Specifically, if we detect the photon back here behind the two-slit barrier, we ask ourselves which slit it went through. The particle, which is one object, can only go through this slit or that slit. It does not make sense to talk about the particle going through two slits at the same time. In the same way, it would not make sense to say that I go through two doors at the same time. I can only go through one door at a time.

DALAI LAMA: But even in this single light, there are quite a lot of particles. A wave itself may be composed of particles, like a water wave. Why are you presenting these as being so totally different?

ANTON ZEILINGER: The reason is—and this an important point—that we can do this experiment with individual photons. I cannot do it here because it would employ a more complicated setup, but it is done in our laboratory all the time. I do the experiment by sending only one photon through at a time and detecting where it lands on my screen. Then a minute later I send the next one through and register where it lands, then the next one, and so on. If you do the experiment with a thousand photons, one photon at a time, you see that these photons have exactly the same distribution as the pattern you saw before, which indicated that light is made of waves. The problem is that you cannot have a picture anymore of a wave made of many particles because you send only one particle through at a time.

With this last exchange we quickly moved to one of the key paradoxes of quantum physics. When light travels through space, it seems to travel as a wave; but when we detect it, light shows up as a particle. The pictures associated with classical physics, pictures such as waves and particles, can be useful under certain conditions, as when using light of high intensity. But modern experimental techniques allow one to work with single photons. Here one encounters paradoxes, and all pictures fail. As the confusing nature of the phenomenon became apparent, the Dalai Lama leaned forward with a look of consternation. He then turned to his translators to discuss these results. Alan Wallace reported on his questions:

ALAN WALLACE: His Holiness was asking whether a single photon travels with a wavelike motion, rippling along through space, and I said no, it goes straight. He asked if a lot of photons together go rippling, and I said that's not true either. So explain this weirdness: Where does the wave come in if all of the photons are going completely straight?

ANTON ZEILINGER: In modern physics we can only talk about a wave going through the slits if we don't ask where the photons are going. If we ask where a photon goes, it may be a straight line. If we don't ask about photons, then we can talk of a wave.

DALAI LAMA: It's rather like throwing the dice for a divination.

ANTON ZEILINGER: Well, there is something to that. The way we look at this problem today is to say you can have a wave picture or a particle picture, depending on which experiment you do. If you do an experiment where you determine the path the particle takes, you use the particle picture, but then you do not think of light as a wave. If you do an experiment like the two-slit experiment and you don't ask where the particle goes, then you can understand it as a wave. But never both at the same time. This is a very deep idea, which was invented by Niels Bohr, a famous Danish physicist. He called it complementarity. You can have different concepts, like particle and wave, which for us exclude each other. We don't know how to make sense of them together. Why does Bohr say these two exclusive ideas are complementary? Because the apparatus that you use to see the wave is different from the apparatus you use to see the path of the particle. The important point which is new in modern physics is that the observer, the experimentalist, decides by choosing the apparatus which one of the two features, particle or wave, is a reality. The observer has a very strong influence on nature, which goes beyond anything in classical physics.

The Role of the Observer in Quantum Mechanics

DALAI LAMA: Am I right that, in terms of the present understanding, nothing can be said about the nature of light independent of any system of measurement whatsoever?

ANTON ZEILINGER: That's right.

DALAI LAMA: It's not clear yet why an observer is involved. What we have so far is the participation of the apparatus. It's very clear why one apparatus as opposed to another has a very direct impact on the perceived nature of light. But where does the observer come in?

ANTON ZEILINGER: This is a question which is debated in physics. My position is that the observer only comes in as the one who decides which experiment to do. He selects the apparatus. In this experiment here I can decide if I want to look at which path the photons take, in which case I use the photon detector and talk about particles. Or I decide not to look at which path the photon takes, in which case I can see the wave pattern. I would say there is no more to it than that. There are some people who claim that there is more observer influence on the experiment.

Here the Dalai Lama makes an important distinction between the participation of a conscious observer in the subjective act of observation and the obvious influence of the apparatus on the light. The direct influence of an observer on an experiment is perhaps the thorniest problem in quantum theory. It is usually called the measurement problem. By contrast, the influence of a physical apparatus on light and thus on the outcome of an experiment may be complex, but it can be described and understood in terms of conventional quantum theory. Anton's position is the most circumspect possible, namely, that the consciousness of the observer only inserts itself into the experiment by choosing the arrangement of the apparatus. No human observation per se has been made, nor—according to Anton—is one required. Other physicists grant observation a more important role. The Dalai Lama will return to this issue later to probe more deeply into the role of consciousness in measurement. The role of the observer is central to Buddhism's philosophy and has a profound relationship to its view of the intrinsically impermanent nature of reality.

DALAI LAMA: Your illustrations are all based upon light. Do these phenomena apply to other things aside from light? Does the wave-particle duality pertain also to sound?

ANTON ZEILINGER: In principle, it does. The problem is that it is very hard to see because the particles in sound waves have a very low energy. But particle effects can actually be seen in sound waves in solid crystals.

I use light because it is the only phenomenon that I can demonstrate here. Similar wave-particle phenomena have been demonstrated for electrons, for heavier particles like neutrons, and even for whole atoms and small molecules. To some extent, they have already been seen for collections of atoms on the order of maybe a few thousand. This leads us to expect that what we are talking about is universal. If you could use the right apparatus, then you would see wave-particle effects for everything. The limiting factor is how large the optics would need to be to see them. If the optics were large enough, you could imagine doing this experiment with billiard balls instead of photons.

The point is that we think that this holds not only for small things but also for large things. It's not a question of size; it's a question of economy because the larger the things become, the more expensive the experiments get. Another problem is that, to see these effects, the quantum phenomenon has to be sufficiently isolated from the environment.

The phenomenon starts and ends with our observation of it. The larger the object becomes, the more difficult it gets to isolate it from the environment. That's a very serious limitation.

From the above exchange we can draw an important conclusion. All wave phenomena—be they sound or light—are also accompanied by particle effects, and likewise all particles (electrons, atoms, molecules . . .) show wavelike effects. Moreover, this ambiguity is universal. As far as we know, there is no size boundary beyond which wave-particle effects disappear. Wave effects indeed become subtler as objects become larger, but with sufficient experimental resources they can always be detected. In other words, physicists now believe that the world is quantum mechanical through and through.

In the next short section we introduce the essential distinction between subjective and objective randomness. We returned to this theme in the afternoon session to understand its implication for Buddhist philosophy.

Randomness in Quantum Mechanics

ANTON ZEILINGER: Maybe I can now address another very important question in quantum physics. I mentioned before that we can do the two-slit experiment with individual photons and observe where they land—say, the first photon lands here, the second will land there, the third one there, and so on. The question now is, Why does a specific particle land at this specific point? As far as we understand today, this individual event is completely random. There is no explanation.

Let me underline the difference between this and classical physics: If I play dice and get a certain number, in classical physics I can at least make a mental picture of what is happening. I can explain why I now get the number 3 because I turned my hand just so, the die rolled that way and hit the surface in a certain way, and so on. Subjectively, I don't have the information, but I can build a chain of reasoning, which in principle would explain it. In classical physics, we call this subjective randomness because I, as the subject, don't know why a particular number comes up. It's just my ignorance. In quantum physics we also have individual random events, but they are objectively random. It's not only that I don't know where the particle will land, but the particle itself does not know. If there were a God, he wouldn't know either. There is no reason why you get a specific result in a specific run of the experiment. This is really the first time in physics that we see events for which we cannot build a chain of reasoning. We can build a reason for the whole pattern: If we collect results for many photons, then we see the striped pattern and we can make our beautiful wave picture. But for the individual particle, there is no way to make a mental model. This has led to big debates, as you can imagine. Some people have even said that what we observe in the individual quantum event is a spontaneous act of creation —something that was created without any prior reason.

DALAI LAMA: Just for clarification—randomness, by definition, precludes any pattern?

ANTON ZEILINGER: Yes, at the level of the single detector. But cumulatively, over time, there is a pattern. This is the paradox. There is an overall pattern, but any individual event is random.

DALAI LAMA: Is it true that the individual events are really random, but when you take them cumulatively then once again causality emerges and you can make a coherent explanation?

ANTON ZEILINGER: Yes, but it is not precise. Because of the individual randomness, we cannot say precisely how many photons will land on which spot. But we can say roughly, and the more we use, the better is our prediction.

The significance of the points raised in this short exchange is hard to overstate. The Dalai Lama was clearly engaged with the issue. In science prior to the advent of quantum physics, the goal had been to give a microscopic causal account for the macroscopic phenomena of our world. The order we see around us was thought to be built on the order of a hidden microscopic world. But here, with the so-called objective randomness of quantum events, that entire enterprise collapses. At the smallest scale, single quantum events are random. How can macroscopic order or pattern emerge from microscopic randomness? When Einstein met this problem, he responded with his famous statement, "God does not throw dice with the universe!" We will return to this issue later.

So far all the effects discussed concern only single particles. A new class of phenomena even more paradoxical than the first arises when we are dealing with two or more quantum particles. To understand the key experiments requires that we first introduce the concept of polarization.

ANTON ZEILINGER: Next I would like to go on to the quantum physics of two or more particles, which also holds deep mysteries for us. But first I need to introduce you to the notion of polarization.

In classical physics, a wave is something that oscillates. A water wave is called a transverse wave because the wave oscillates transverse to the

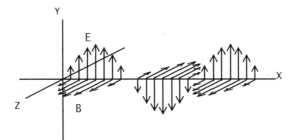

Figure 1.3 If light is polarized vertically, its electric field oscillates vertically. The electric field (E) oscillates up and down parallel to the Y-axis. The magnetic field (B) vibrates perpendicular to the electric field, and so is parallel to the Z-axis. The wave is traveling to the right along the X-axis. Polarization is given by the electric field direction, and so we say it is "vertically polarized."

direction of propagation. Light is also a transverse wave of electric and magnetic fields, which means that if a light beam propagates in one direction, its electric and magnetic fields oscillate transverse to that direction. The oscillation can occur in different orientations, but in all cases the oscillation is transverse to the direction of propagation. Each oscillation direction corresponds to a different form of polarization. I have a very simple set up here: a laser beam and a polarizer, which only lets light through that oscillates in one particular orientation. No other light goes through. And here I have a second polarizer.

Anton directed the laser beam through the two polarizers, placed one after the other (see figure 1.4). Unpolarized laser light enters at the left; that is, the electric field oscillates in random directions perpendicular to the direction of propagation. After passing through the first polarizer, whose transmission axis is vertical, the electric field is linearly polarized vertically.

ANTON ZEILINGER: You can see the red spot of the laser light because the two polarizers are now parallel to each other. What goes through the first also goes through the second. Now I rotate one of them, and they are orthogonal. The light oscillating in a particular direction goes through the first polarizer. But then it meets the second polarizer, which only allows light oscillating in a perpendicular direction to the first, so nothing goes through.

Experiment and Paradox in Quantum Physics

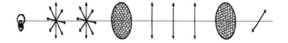

Figure 1.4 The phenomenon of polarization of
light. From the light source we have unpolarized
light whose electric field oscillates in all possible directions transverse to the line of propagation. After
passage through the first polarizer, the light is polarized in one direction only. This can be tested by
using a second polarizer, which can be rotated. If the
two polarizers are parallel, all light passes the second polarizer. If they are oriented at right angles, no
light passes the second polarizer.

Nonlocality and Entanglement

ANTON ZEILINGER: Now I want to introduce an experiment that is too
complicated to bring here, but if Your Holiness ever comes to Innsbruck,
I would be pleased to show it to you in the laboratory. We can do it here
as a *gedanken* experiment—a thought experiment that is conducted in
your head but which follows the rules of physics exactly. These are the
cheapest experiments. Thought experiments have been very important in
the development of physics in the twentieth century because the consequences of relativity theory and quantum mechanics were so strange that
people could not do real experiments in the beginning. This experiment
has in fact been done in the lab many times, in ever more refined ways,
but it started out as a thought experiment. The first ideas in this direction
were presented by Einstein in a famous 1935 paper, the so-called Einstein-
Podolsky-Rosen (EPR) Paradox. The version of the experiment using polarization was invented by David Bohm in 1952.

What we have here is a simple source that sends out two photons simultaneously in different directions. [*See figure 1.5.*] We don't need to
know the inner workings of the source. Then we measure the polarization of each of the two photons. We put a polarizer in the path of each
photon, and we place a photon detector behind each polarizer. Then we
simply look for coincidences. Sometimes only one detector clicks to indicate a photon and sometimes both click. A coincidence happens if we
register a photon behind each of the polarizers.

The experimental observations are very basic. The first observation
is that whenever the two polarizers are oriented parallel to each other,

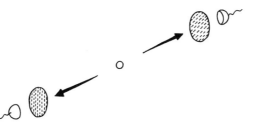

Figure 1.5 Experiment on the correlation between two photons (Einstein-Podolsky-Rosen experiment). A source emits pairs of photons. Each one is subject to a polarizer. One detector on each side serves to find out whether the photon passed its polarizer or not. One then investigates how often both photons pass their respective polarizer—which depends on the relative orientation of the two polarizers.

you get coincidences: For each photon registered on one side, another is registered on the other side.

DALAI LAMA: Does this happen invariably?

ANTON ZEILINGER: You have to account for the fact that your detector is not 100 percent efficient, but if you take that into account, then, yes: If the polarizers are parallel, either both photons are detected or neither are detected.

Now we can start to make a picture to understand what this means. The most natural picture is that these two photons start their travel with the same polarization or direction of oscillation. If they have the same polarization, then if one goes through, the other will also go through. More generally speaking, each photon has properties—a set of rules or a list of instructions that defines what to do when it meets a polarizer. If the polarizer is oriented a particular way, the photon goes through. If the polarizer is in another position, the photon doesn't go through. You can explain why both photons behave the same way by assuming that both have the same set of instructions—the same feature, whether it is polarization or something more complicated.

If you use this model, then you arrive at certain predictions for the number of coincidences when the two polarizers are not parallel. If the two polarizers are neither parallel nor orthogonal but somewhere in between, you will get coincidences sometimes but not all the time. The interesting point, which was John Bell's great discovery, is that for certain orientations, the model predicts a lower number of coincidences than we observe. It turns out that there is a conflict between experimental observation and any model that follows the same ideas—not just the model I have been describing. Regardless of how clever we are, no model can explain the observation on the basis of the properties of the particles taken separately. It is absolutely striking.

What do we conclude from this? Whether a photon goes through the polarizer is not determined by its properties. Each photon has a 50 percent chance of going through; that is, it is perfectly random. The mystery is that each of the two photons performs randomly when considered separately; but when the polarizers are aligned parallel, then both always perform the same. As an illustration, suppose you have two dice, and you give one to your friend. He goes off to a star far away. And at some time, you throw a die and your friend throws a die. It turns out that even though both of them are completely random, they always match. That is the mystery. How can two random processes give the same result?

Physicists describe this as *nonlocality*. The word implies that the measurement obtained for the photon on one side depends not only on the orientation of the polarizer on that side but also on the orientation of the polarizer on the other side. Conversely, the definition of *locality* is that what we observe here and now does not depend on what somebody does far away at the same time. Nonlocality is one way to understand what is going on, although *understand* may not be the right word. Nonlocality is a way to describe the situation, but it is not an explanation. Einstein introduced this concept, but he did not like it.

DALAI LAMA: When you talk about dependency here—that what happens on one side seems to be dependent on what happens on the other side—you are not talking in terms of causal dependence, are you?

ANTON ZEILINGER: That is a very deep question and an issue of debate. People have tried to make causal models to explain this. I personally would say no, but that is a matter of taste.

DALAI LAMA: Is it true that in modern physics there is generally not an acceptance of simultaneous causality?

ANTON ZEILINGER: That's right. This was stated by Einstein most clearly. Any cause has to propagate with the speed of light at most.

The name we use to describe the above type of nonlocal connectedness of two particles is *entanglement*. The idea is that under certain circumstances two particles remain one system even if they are separated by a very large distance. They are not really separated in a deep sense. It becomes even more strange, and very complicated, if we talk about three particles. We can keep going and talk about four or five or six particles. It never ends. But time does not permit me to go into this; I have to stop at two.

DALAI LAMA: Are you implying that the entire universe is internally entangled?

ANTON ZEILINGER: That is a nice idea, but I would not want to take a position on that because, as an experimentalist, I would not know how to prove it. The intellectual and philosophical hero of this field was Niels Bohr, who made a very wise statement: "No phenomenon is a phenomenon until it is an observed phenomenon." In other words, we should not talk about a phenomenon unless we observe it in a real experiment. [Laughter.]

DALAI LAMA: Someone would probably have to live very long to prove it, to be able to see the whole.

ANTON ZEILINGER: Exactly, and he would need a lot of money.

DALAI LAMA: Perhaps we could ask the Pentagon. . . .

ANTON ZEILINGER: Then we would need an enemy, and there is no enemy.

Let's conclude with this picture of the Borromean rings, which I really like as a description of entanglement for three particles. On the entrance to the castle of the Borromean family in northern Italy, you see their coat of arms with these rings. If you look closely, you see that the three rings are connected in such a way that if you remove one, the other two also are no longer connected. This is the situation in the entanglement of many particles. They are all connected, but if you take one out, the others are not connected anymore. It is quite interesting. [*See figure 1.6.*]

Figure 1.6 Two different ways in which three rings can be interconnected. In the case of Hopf Rings (left), any two remaining rings remain connected to each other if one of the rings is taken out. In the case of the Borromean rings (right), the other two rings are not connected if one ring is removed. Both situations can be verified by one and the same three-particle entangled state. Whether or not they behave as Hopf rings or Borromean rings depends on the kind of measurement performed on the three particles.

Erwin Schroedinger considered the concept of entanglement, introduced above by Anton, to be the defining feature of quantum mechanics. Entanglement fundamentally challenges our conventional conception of objects as entities that have a persistent and unambiguous set of attributes. Quantum attributes can be ambiguous and nonlocal. This raises deep philosophical issues about the very nature of objects considered quantum mechanically.

Causality in Quantum Mechanics

I asked if the Dalai Lama had any questions. He deferred to the translators, who were Tibetan Buddhist scholars in their own right, and asked them to put forward their own questions. They returned to the question of causality and randomness. In particular, they pointed out that whereas some causes may be hidden from initial view and are therefore unknown to us, this does not necessarily mean that they do not exist. Indeed, Einstein held a similar view, which goes under the name of hidden variable theory.

ALAN WALLACE: For these quantum events, you have not found a cause that determines which slit the photon goes through. But you've not demonstrated that there are no causes. It's not clear to me why you don't simply say, "We don't know," and leave it at that, rather than asking us to throw away thousands of years of belief in causality. You could just say that all theories posited on local hidden variables don't work, and rest there. You haven't demonstrated to us at all how you have recognized that there aren't any causes.

THUBTEN JINPA: This is important because Buddhist analysis makes a distinction between that which is not found and that which is found not to exist. There is a big logical difference between the two.

ANTON ZEILINGER: I think part of the problem is that we seem to have different notions of what we mean by causes. But first, I would like to make one little correction. I certainly did not come here to request that anybody give up thousands of years of tradition—just the opposite: I want to learn from that tradition. In unifying our views, it might very well be that both sides have to give something up.

On the specific question you raised, we have not proved that there are no causes. If we speak of causes, we use a very specific notion of cause within the specific experiment. In the two-slit experiment, if I put a detector behind each slit, then there are only two possibilities. Sometimes one will click, sometimes the other, and it looks completely random. Fine—we have not found any cause. But we have also done experiments on entangled states where we can show that causes cannot exist insofar as they are local. That's an important point. We have learned from relativity theory that any influence can move only with the speed of light, not faster. We have confirmed that if causes exist, they cannot be local and they do not explain why this detector fired rather than that one. That much we know. There are other possibilities. One possibility, for example, would be nonlocal causes.

At this point in time one is faced with a personal choice: what to believe in this situation. We have now seen seventy years of quantum physics and many people trying to understand what's happening on the basis of old notions, and none of them have solved the problems. I like to take the most radical view possible: that the next plateau must be even more strange and weird than what we have now. I take as a working hypothesis that these things are completely random and there is no explanation. I think that is very likely to be true, although at present that's a matter of judgment. But I want to know what that really tells me about the world, myself included.

Can I add just one little point? We know that it is not information carried separately by the two particles. When we talk about nonlocal action, we are saying, when this particle is measured, some information instantly travels over there and tells the other one. Piet has reminded me that there is another very important debate connecting nonlocality with the relativity of time. If the two photons are registered at the same time relative to one location, then I could imagine somebody else coming by on a train, seeing one happen earlier than the other. He would say that one is the cause and the other is the effect. Another person going in the opposite direction would see the reverse cause and effect. Aside from the point that we have not yet been able to identify a nonlocal cause, we run into the further problem that we would not know which is the cause or the effect.

ARTHUR ZAJONC: I would like to present a view that might be more consistent with Buddhism and is still consistent with modern physics. David Bohm's understanding of nonlocal causality would support hidden causes or hidden variables. In the two-slit experiment, for example, he would say that each specific photon has a predictable trajectory. He can compute the trajectory and tell you ahead of time which photon will go where. The one problem with his theory is that every calculation must have an initial state. If you tell me the exact initial placement of the photon, then I can tell you the rest of the story. From the standpoint of normal quantum mechanics, you can't actually determine this initial placement. So even in Bohm's interpretation of quantum mechanics we have the randomness we see experimentally. In principle one could derive the initial state as the end product of a previous evolution. So the beginning gets pushed back a little further and a little further at each step. If you accept the concept that there is no beginning, then the entire story is consistent. The problem arises when you have to begin. Bohm's theory is an example of a nonlocal quantum theory, of the type Anton was describing. Bohm would say there is really an underlying cause for the apparent randomness, and it is no different from the usual source of randomness, namely, ignorance. In this case it is ignorance of the initial state of the photon. Other physicists would criticize Bohm's theory because it requires the initial conditions, which cannot be provided. But maybe from the Buddhist standpoint there are no initial conditions, and you can trace causes back infinitely.

Entanglement Broken

The question then arose as to when entanglement of particles with each other ceases.

ANTON ZEILINGER: When is entanglement broken? Entanglement is broken when one of the two particles interacts with an outside system such as a detector. In other words, once the detection is made, there is no entanglement for future observations. The first observation breaks the entanglement.

ARTHUR ZAJONC: A caveat should probably be added here. Some people say that the detector is a quantum system itself. Therefore, what takes place is not disentanglement but an increasing entanglement of one quantum system with a much more complex quantum system, that is, with the detector. The problem goes on forever. This leads to the view that Your Holiness asked about, the view in which the whole world becomes entangled. A more modest suggestion is that beyond some boundary the detection becomes irreversible; it becomes too complicated to properly describe quantum mechanically. The detection event entangles a huge number of atomic systems, which effectively produces a classical disentangled result. But many physicists, John Bell in particular, would say this is not true. For all practical purposes, it is true. But in principle, philosophically and scientifically, it's not really true that entanglement is broken. The detection increases the scope of the entanglement rather than breaking it.

ANTON ZEILINGER: That is an important point and it underlines why the Copenhagen interpretation is so strong. I said observation breaks the entanglement, but I did not say what constitutes observation. Observation is a primary unexplained notion in the Copenhagen interpretation. Observation is when we take notice of the features of the classical apparatus, and then we have no entanglement anymore. If we try to keep using quantum physics for the pieces of the apparatus itself, then we run into all kinds of problems, like superposition of the apparatus or the revered Schroedinger's cat paradox. Then there is the Wigner's friend paradox. Eugene Wigner was a famous physicist who said, "Well, the detection apparatus is entangled, and then my friend looking at the apparatus is entangled. I alone am not entangled, because my brain does not realize entanglement." The paradox is: Where does this chain break? This is not a problem if you take observation as a primary notion. Then there is nothing to be explained and that's it.

It is clear that we do not experience the world in the bizarre quantum mechanical states that theory tells us exist. Schroedinger's cat is perhaps the most blatant example of the absurdities one gets into if we don't break entanglement at some point. Schroedinger showed that if taken to its logical extreme, entanglement can lead to a cat's being both alive and dead at the same time. This is Schroedinger's cat paradox, referred to by Anton. Zeilinger's viewpoint concerning this and other mysteries of quantum physics is similar to that advanced by the Danish physicist Niels Bohr and his colleagues, which has come to be called the Copenhagen interpretation. I cannot summarize this subtle philosophical view in a few lines, but it is important in the context of Anton's remarks to say that in the Copenhagen interpretation observation is not explained in terms of a detailed physical model. It is, as Anton says, "a primary unexplained notion."

This is important because as soon as one offers a physical model then that model can be analyzed quantum mechanically, and all the problems associated with measurement again rear their heads. One alternative is to invoke observation (without explaining what it is) as that which breaks quantum entanglement and leads to the classical readings of our scientific instruments. Another alternative, one advanced by the famous physicist Eugene Wigner, seeks to break the infinite web of entanglement through human consciousness. In this view, human cognition somehow breaks the chain of entanglement and produces classical reality when the nonphysical mind of a human observer gets involved. Someone then perversely asked what would happen if Wigner's friend observed the cat? Would the paradox collapse or persist—since from Wigner's viewpoint the friend is no different than the cat or the rest of the apparatus? Whose consciousness counts: the cat's, the friend's, or only Wigner's?

During our first morning discussions, many of the major issues of quantum physics emerged: wave-particle duality, objective randomness of quantum events, multiparticle entanglement, nonlocality, the transition to macroscopic effects, and the measurement problem. Each of these features of quantum mechanics is supported by experiment, and each requires a profound reconceptualization of our world and even of the role we play as observers. We had only begun the conversation, but already profound philosophical questions were before us. These and others like them would occupy us in the days ahead.

2

Philosophical Reflections on Quantum Realities

The facts of quantum physics have puzzled physicists and philosophers for the last century. Even those who invented the theoretical language used to describe quantum phenomena, such as Erwin Schroedinger, did not see how the mathematical theory was to be understood. Niels Bohr had suggested that theory could no longer provide a picture of reality and that we therefore must be satisfied without one. Others, like Schroedinger and Einstein, remained convinced that eventually new insights would provide a more satisfactory account than the purely formal one given through the mathematics of quantum mechanics. Schroedinger put it this way:

> *A widely accepted school of thought [Bohr's] maintains that an objective picture of reality—in any traditional meaning of that term—cannot exist at all. Only the optimists among us (and I consider myself one of them) look upon this view as a philosophical extravagance born of despair in the face of a grave crisis. We hope that the fluctuations of concepts and opinions only indicate a violent process of transformation which in the end will lead to something better than the mess of formulas that today surrounds our subject.[1]*

Is there a picture of reality that is objectively true? Was Bohr correct in abandoning all hope for such a picture of reality? Or do we need to think more subtly—yes, giving up a fixed picture but allowing for a more flexible and appropriate form of understanding, a position between the realism of Schroedinger and the positivism of Bohr?

Like Schroedinger, Einstein, and Bohr, we in Dharamsala were confronting the fundamental philosophical issues raised by modern physics, but now we were doing so within the enlarged framework achieved by including Buddhist philosophy, as well as more familiar Western sources, in our inquiry. Schroedinger had been a careful reader of the Bhagavad-Gita and a student of Sanskrit. Like him, we were interested in extending our range of thinking to include non-Western sources in the hope that these added intellectual tools would help us to gain greater clarity into our own Western scientific and philosophical problems. After all, Buddhist philosophy has struggled for centuries with the issues of ontology and epistemology in ways that bear directly on the standing of the pictures we form of reality. Where science has traditionally adopted a straightforward realism, Buddhism has been far more circumspect about the ultimate nature of reality. According to it, our experience of the world, including that given through experimentation and scientific theory, concerns what it terms "conventional reality." According to Buddhism, a deep philosophical analysis of reality reveals its ultimate emptiness. Buddhism adopts a more fluid and phenomenological view of reality. Reality is considered to be a series of momentary phenomenal events. Moreover, these phenomenal events do not originate purely from the side of the external world alone but rather are contingent on a complex causal nexus that includes the mind. This is the venerable Buddhist doctrine of dependent arising. As the conversation develops between scientists and the Dalai Lama, it will be important to remember the very different views of reality the two parties traditionally hold and to notice the ways in which quantum mechanics helps to bring them closer together.

In a way, Tu Weiming embodied for us the blend of East and West we sought. Born in China and educated in both Taiwan and North America, Weiming understood better than most the contrasting intellectual and spiritual traditions of Asia and the West. His comments each day opened avenues for possible synthesis and collaboration. His presence was also an important reminder that no matter how tragic the history and politics between Tibet and China have been, as human beings we can rise above the fray in our common striving for insight and the desire to relieve suffering. For these reasons, His Holiness was especially pleased that Tu Weiming was able to participate.

Referring to the morning's introduction from Anton, Weiming and I began the afternoon's conversation with the Dalai Lama. Weiming gave eloquent expression to the challenge of the new science for our view of the world and the potential benefits of collaboration between different disciplines and traditions—East and West. Moreover, he expressed his interest in a fuller participatory role for the mind of the scientist or observer than that supposed by Anton. The Dalai Lama responded by discoursing on the "four modes of analysis," the nature of emptiness, and dependent origination in Buddhist thought as they might pertain to the issues raised by quantum mechanics.

Objective Randomness and New Philosophical Views

ARTHUR ZAJONC: This morning Your Holiness gained an impression of how quantum experiments and their interpretation challenge our normal way of understanding the world in which we live. When these experiments and theories were first presented at the turn of this century, from around 1900 to 1920, the scientific community was profoundly shocked. People simply could not understand how these effects and the theory that explained them could be true. Gradually, over the course of decades, people began to understand the impact and the full consequences of these ideas. But it's important to say that scientists, Anton and myself included, are still amazed by these experiments. It's not that they somehow became routine. They still have an extraordinary impact on the way we see the world.

One of the areas we wanted to discuss further this afternoon was the question of objective randomness. We said that subjective randomness is based on ignorance: If we are ignorant of the causes of something, then it appears to be random or chaotic. Once we understand the true causes, then patterns are recognized. In quantum mechanics we meet a domain of phenomena characterized by randomness, but they appear, to the best of our judgment, to be objectively random. That is to say, there is no hidden cause, even on an extremely subtle level. More knowledge does not reveal the true cause, but rather nature itself seems to have this characteristic of being random. It happens at a very deep level. In the normal course of daily events, things still proceed in a causal and predictable way. But at the deep level of quantum reality, nature appears to be objectively random. We would like to discuss this mystery further in relation to Buddhism. Is there any place for objective randomness in Buddhism? What is the nature of the relationship between cause and effect, between ignorance and randomness? But first I would like to invite Weiming to make some opening comments.

TU WEIMING: I am very privileged to be here, both in the presence of His Holiness and in the presence of some outstanding scientists. I represent the lowest common denominator in this group—possibly in the sense of having a "beginner's mind," though I'm not sure, but certainly coming from ignorance.

My intuitive response to the very fascinating notion of objective randomness is to hope that it's not true. I hope that this is a stage of development in our knowledge and that eventually we will have a much better appreciation of the ultimate reality of the world. My wish is that these things happen not simply randomly, and especially not in the objective sense, but because of all kinds of subjective conditions, including our conceptual apparatus and our ethos—and that with more subtle ways of understanding we will have a different perception. But this is wishful thinking, of course. It is totally rejected by the incredible results of the instruments we used this morning. The instruments show us, first of all, a concept which is very contradictory to my own ordinary conception: a complementarity that involves two immeasurably different and incompatible explanatory models.

That complementarity evoked in my mind a very classical Taoist idea about using language as an instrument to capture meaning. The metaphor may be misleading, but the notion is that using language to understand meaning is like using a net to catch fish. Often people who are not trained are confused, identifying the net with the fish, confusing language with meaning. The instrument that you use to catch the fish defines your conception of what the fish is, and it becomes instrumentalized in the wrong way. But there is no way of catching the fish other than with the net. This is the only instrument that we have, and therefore the instrument becomes a constitutive part. Language becomes a constitutive part of the meaning we try to capture. No matter how effectively we try to use the language, our meaning is being conditioned and shaped by this particular procedure.

There are three dimensions in this analogy: the observer or the person, the instrument used for understanding, and the objective phenomenon that we try to understand or conceptualize. My sense is that, at the present stage of development, there is a conflict in interpreting what we have discovered. The conflict of interpretation is linked to two very different perceptions.

One perception is the deterministic model we hold: No matter how random the situation seems, if we work hard enough to broaden our knowledge base, we will reach a higher level of understanding. The investigation that we are focusing on is only part of a much larger reality, and if we broaden our scope, eventually we will find a comprehensive picture which is not random.

The other position is more radical, and I think an increasing number of scientists accept it. It's not simply that it's difficult for us to find the causality, but we come to the realization that there is no causality. The belief that there is some kind of causality—the lack of acceptance of the randomness—is fundamentally flawed, and the more we try, the less successful we will become.

These two positions have consequences. One position assumes that many of our old beliefs in stability—in the substantial, predictable continuity of the world—may still be rescued. The other is more radical. It says we have to find a radically new theory—not just a new instrument, but a totally different view of the world. If we cannot find that and we remain attached to our old habits and vision, no matter how hard we try, we will never be able to understand the world.

If we take the second, more radical view, we find ourselves at a very critical stage. Not only do we have to be very skeptical of all received knowledge about the ultimate reality of the natural world, but we need to be totally open to all kinds of possibilities and new ways of doing things. And we have to be excited by the new knowledge that has become part of the scientific community and may eventually become part of the larger community. In that frame of mind, I would like to ask for the wisdom from your own tradition, especially concerning dependent origination and emptiness. How do you react to these quantum phenomena?

My hope is that a radically different conception of the universe will arise, not only the totally new view imagined now by some of the scientists, but maybe one drawing on comparative cultural studies in surprising ways. We shouldn't rely on the power of the modern West as the only paradigm for explaining what's going on in the world. This paradigm emerged in the eighteenth century as part of the Enlightenment movement and reached a climax in the nineteenth and twentieth centuries. It places a great deal of emphasis on the importance of rationality: using analytical methods to dissect things into their smallest possible parts; quantifying precisely; trying, hopefully, to explain things with a mathematical model; and holding the ideal of a disinterested observer who is not an integral part of the process. Many of the great accomplishments in modern Western science became highly problematic because of the new developments in physics. We are at a stage where new knowledge will have to come from a much broader collaborative effort. That collaborative effort may involve people from many different disciplines and different traditions but with a precision that has been advanced by science. This morning's discussion depended on the irreducible importance of the empirical method in trying to find out exactly what's going on. Quantum physics is not a romantic assertion; it's very precise, and yet it opens new possibilities to the imagination. It is quite possible that this enterprise of a dialogue between civilizations will reintroduce some modes of thought that were rejected by scientific minds beginning in the eighteenth and nineteenth centuries—modes that were rejected as unscientific because they related to religion or metaphysics. It may be time now to bring many other kinds of vision—artistic, religious, spiritual—to bear on these very important questions.

I would like to conclude with one thought concerning the scientist or observer and what is the vision of the scientist or observer when confronted with this major challenge. This morning's discussion made clear that among these three things—the object that we are trying to understand, the instruments with which we might understand it, and the observer as subject—the instruments themselves are critical. The scientist affects the outcome by choosing instrument A or instrument B. This is important, but it is not participation. Whether or not the scientist has cultivated a higher level of spiritual understanding is basically inconsequential to what is going on. But perhaps we are now at a time, in our search for new explanatory models, when all the qualities of a scientist as an evolving and self-cultivated human being may be relevant for this broader discussion.

Buddhist Analysis

DALAI LAMA: In Buddhism we identify four modes of analysis. First there is the one that Anton has exemplified: setting up a situation, investigating it, asking one question after another, collecting evidence, and then coming up with a coherent explanation. The second mode of analysis is called *chunye* in Tibetan. In this you just simply have to say, "That's the way it is." It is the nature of the phenomenon and there is no further explanation. For example, if you should ask why a certain frequency of light appears blue as opposed to yellow, there is probably some point at which you would say that's just the way it is. When a certain frequency strikes the retina it sets up electrochemical events in the visual cortex, but finally you just happen to see blue and there may not be any further explanation. The third mode of analysis explains an event or phenomenon in terms of its functions, and the fourth mode pertains to the capacity or potential of a given phenomenon. But it may be the second mode that is especially relevant here.

ARTHUR ZAJONC: You mean, it might be that objective randomness is an instance when the analysis comes to a kind of threshold and then these events become probabilistic, without any other explanatory reasons?

DALAI LAMA: In some cases it may be because of the current limitation of our knowledge. In some it may have to do with the limitations of the instruments. But in some cases, it could be an objective feature of reality for which there cannot be any explanation.

Considering the possible relevance of the Buddhist philosophy of dependent origination for understanding objective randomness, my personal feeling right now is that there is probably no connection. The principle of dependent origination is based on definite causal relationships between different events or facts. The dependence may also be understood in terms of the relationship between parts and the whole that they constitute; or it may relate to our means of knowing. Although the notion of dependent origination is based on known relationships between identifiable events or facts, Buddhists would argue that the dependence does not entail that these interacting events or facts have some kind of intrinsic, objective reality in and of themselves, but rather that this absence, or emptiness, of independent existence is at the heart of their existence. Their existence and reality can make sense only within the context of interrelationships and interconnectedness. Insofar as certain experiments in quantum physics point toward an understanding of the nonsubstantiality of material things, then perhaps there is a meaningful parallel with the Buddhist concept of emptiness.

On this theme of emptiness, let me pick up the strand that Weiming brought before us—the relationship of a term to its referent, which is like the net and the fish. Both the term and the object which is its referent exist. But when you seek out the nature of that referent, the entity itself, under critical analysis, you don't find what its nature is inherently. Is the conclusion therefore that it doesn't exist? No, that's a false conclusion: It does exist. How can you have those two statements together—that it exists, but when you look for it you don't find it? It does exist, but it exists by the power of the nominal or verbal designation of it. It's also not the case that just by saying something you bring it into reality. It's not that simple, for sure, because if it is mere construct then it is only imagination. But it is true that there is nothing that has its own intrinsic nature independent of the verbal designation of it. It exists by the power of the verbal designation, and yet, having said that, it's not whimsical.

TU WEIMING: I think that the minimalist interpretation of the experiment is that objective randomness and nonlocality are perceptions derived from immediate experimental, experiential evidence. That is the only thing Anton is willing to say. There is no ontological commitment to a theory. Some other people may have ontological commitments to a larger theory, and they may come up with broader interpretations. To say that nothing comes into being without the process of dependent origination—that nothing comes into being inherently—that is a very powerful ontological commitment. I don't think the experiment so far has conclusively rejected that notion.

DALAI LAMA: In my discussions over the years with scientists, I have begun to sense that much of the scientific approach to analyzing the nature of reality, particularly at the level of particle physics, seems to concentrate more on what Buddhists would call the process of negation—what it is not. As we look for the contingent parts of material phenomena, we begin to realize that there is no substantial reality there. But there seems to be perhaps less emphasis placed on the dimension of existence: In what sense *do* phenomena exist? It is in this context that the Buddhist theory of dependent origination is presented. We try to understand the dimension of existence—what kind of identity emerges in the aftermath of a deconstruction through the process of negation. Of course, within Buddhism there are different understandings of what is meant by dependent origination.

In the above conversation we have entered quickly and deeply into Buddhist philosophy. Very briefly, the Dalai Lama's treatment of emptiness and dependent origination generally follows that of Nagarjuna, an eminent second-century C.E. Indian Buddhist philosopher. Nagarjuna's most important work elaborated the so-called centrist or middle way, known within Buddhism as Madhyamika. In the words of the contemporary Buddhist scholar Paul Williams, Madhyamika is "an attempt systematically to set forth, demonstrate, and defend an understanding of the way things really are."[2] Nagarjuna advanced his position by criticizing the shortcomings of both the realist position of Abhidharma scholars and the idealism of the Cittamatra, or mind-only school. Madhyamika is proposed as a middle way between these two extreme positions. It is a subtle and complex position, and it is widely held in Tibetan Buddhism to be the most advanced philosophical treatment of questions about the nature of reality.

Central to Madhyamika is its view concerning conventional versus ultimate existence and its account of phenomenal experience in terms of dependent origination. Buddhist philosophers speak of two contributing factors to phenomenal experience. On the one hand is the contribution from the "side of mind," and on the other hand is the object it is said to contribute from "its own side." Dependent origination sees conventional ex-

istence as arising from both sides, with no underlying absolute reality to ground the whole. By contrast, Abhidharma realists recognize that our immediate impressions are not fundamental, and they search beyond outer appearances to discover the real, enduring dharmas, *or essential constituents out of which conventional reality is composed. These* dharmas *are unconditioned by anything beyond themselves. They are absolute and so said to "be given from their own side" only.*

Another related theme is that of participation, as pointed to by Weiming in his opening remarks. In what follows, the Dalai Lama presses Anton and me to explain our view of the role played by the scientist not merely as passive perceiver but also as participating knower. Has science properly understood the role of conceptual and verbal designation as it attempts to describe quantum reality?

Participatory Observation and Dependent Origination

PIET HUT: It was very interesting for me that Anton used the term *objective randomness* when he said in his presentation that our understanding of reality is insubstantial and spontaneous. I understand that you like the notion of insubstantiality, as it is presented by Nagarjuna or in the theory of dependent origination, but you do not like the notion of spontaneous occurrences. I am very curious why you do not like the notion of spontaneity.

DALAI LAMA: It's not so much a matter of liking one or not liking the other, but simply that I find some parallels between the thought of Nagarjuna and the quantum mechanical views on the lack of substantiality of photons. As far as objective randomness goes, it's not that I don't like it, but I feel that they probably will find some reason for it as research continues. It could be the case that in certain realms one may find pure randomness as an objective feature of reality that has no explanation whatsoever. But I didn't comment on this because there is little parallel between this and the Madhyamika notion of emptiness.

ARTHUR ZAJONC: Maybe I can add a related comment that also connects with something that Weiming mentioned. In physics, as in Buddhist philosophy, there are different schools. Everyone agrees on the facts of the experiment and begins with the same experimental data. Everyone also accepts the basic mathematical theory of quantum mechanics, but there are different interpretations of this theory. What Anton has presented to Your Holiness can be characterized as the minimalist interpretation, where the fewest possible assumptions or preconceptions are included in the explanation. Anton recognizes the phenomena, the patterns, even the mathematics, but everything else he leaves aside in the hope that some new insight will arise.

This approach has a great tradition, going back to Niels Bohr, although it is a minority position in the quantum physics community. It's a very distinguished minority, but it's a minority. In other words, there are many other schools. Some physicists try to give a very detailed causal account of how the photon travels a specific path and then falls on the detector and causes a specific event. They describe every moment in the history of the photon. David Bohm was one such individual. He felt that Niels Bohr, through the force of his personality, had compelled the physicists of the day to share his belief. These are two extreme positions, the minimalist and the very detailed causal account, and there are many intermediate positions and variations, which we need not go into.

However, all of these models, to the best of my knowledge, somehow include the fundamental point of randomness that Anton emphasized. This has to be part of the model. The other necessary element is what he called entanglement, or nonlocality, where two particles arising out of a single source have a very unusual, nonlocal nature. David Bohm includes both of these in his theory but in a way that remains puzzling. He assigns the randomness to the preparation of the photon. Once it is randomly prepared, it follows a very clear path, but why it starts in one place or another is pure chance. Bohm also includes nonlocality in the theory, but he explains it through a special kind of force he calls the quantum potential. The two entangled objects are embedded in a quantum force, so when you move one, it moves the other. It's a quantum causality rather than a signal that propagates from one part of the universe to the other. So Bohm gives a causal explanation of why the two particles are correlated, but it is a nonlocal causality that conflicts with common sense. The important point is that you always need to account for the same features of randomness and nonlocality, regardless of which school of interpretation you come from.

Another very important point I would like to emphasize is the way that order arises out of the coincidences between the two detectors. If you only look at one detector and pay no attention to the other, you just see random clicks. But the remarkable thing is that when you stand back and observe both, then an order appears.

DALAI LAMA: Perhaps it's from the observer's point of view that the order arises. It is the observer who sees the two events together.

ARTHUR ZAJONC: The character of the attention brought to the situation affects whether the order arises as a phenomenon. If the attention is limited to one detector, then it appears to be random.

ANTON ZEILINGER: We can take the observer out of this. You can do the experiment in such a way that each detector is very far from the source. A technician at one just writes down his results and the time when it happens: At five o'clock the photon came through. Another technician far away writes down the data she gets. They both have lists, which taken alone make no sense at all. When they meet and compare their lists, they find out there is this interesting connection between them.

DALAI LAMA: But there is still an observer. The very orderliness of it is dependent upon the observer noting those two phenomena. It doesn't matter, does it, whether it is one or two people? The fact is that an observer is engaged here and looking at two things.

ANTON ZEILINGER: But that observation happens much later, right? You put your records into the safe, and a year later you take them out and look at them.

DALAI LAMA (*responding with a laugh*): The observer is not necessarily a perceiver but a designator, a knower. The very notion of order does not exist without the designator being there looking for it and noting and superimposing upon these phenomena: "Ah! There is order." If you don't have a designator, a knower in some sense of the term, you are back to disorder.

ANTON ZEILINGER: As long as the knowers are only paying attention to a part, then they miss the order.

DALAI LAMA: Do they miss it, or does it not exist?

ARTHUR ZAJONC: This is a task for philosophy. I would say it exists. Anton would probably say that it does not exist, but I should not speak for him.

DALAI LAMA: Is the very notion of order something that is dependent on an observer, a knower? Is it an absolute, a purely objective feature? Is order something that is independent of any specific subject?

ARTHUR ZAJONC: Well, here you come to actualization versus the potential to be known. You sometimes have more than two categories. You can have the presence of something. You can have the absence of something. But, following Aristotle, it can also be potentially present. I would say that order is potentially present. It only becomes actually known, of course, through a conscious mind. In that sense, I would personally say, yes, there is an ordered universe.

ANTON ZEILINGER (*jumping in with a smile on his bearded face*): Can I first disagree with what you said about my opinion? Then can I ask a question in return? Suppose there are two monkeys playing with the polarizers, and there is an automatic registration of the time at which each polarizer was set in which way and whether a photon was detected. And then much later these records are brought together and you see the order. Now who brings the order in there? Was it the monkeys playing with the apparatus? Or the person bringing it together later? Where does the order come from? In some sense, it is already there in the data. The final data will show the strange quantum correlation.

GEORGE GREENSTEIN: Can I make a point here? We're having a discussion about order, but that's not the essential point here. Let me describe a slightly different experiment. You have the source and the two detectors. You add a bell that rings only if the detectors click simultaneously; if they do not click simultaneously, it doesn't ring. Now the only question is, Did the bell ring? This has nothing to do with order or disorder; it has to do with the ringing of a bell. The same mysterious bell ringings will occur, and that's the thing to focus on. Order or disorder is some judgment we make, but the ringing of the bell is a separate question.

DALAI LAMA: But the ringing of the bell is associated with the polarization. That is where the whole issue of order comes in. It's not just the bell ringing. The fact that there is a relationship between the polarization and the ringing of the bell involves somebody evaluating the situation and then projecting upon that: "Ah! Order."

GEORGE GREENSTEIN: No, there is just a bell ringing.

ARTHUR ZAJONC: Yes, but it's seen in correlation with parallel polarizers.

GEORGE GREENSTEIN: Yeah, but that can be done by a machine. Anton could build a trivial little machine in ten minutes, coupling the two detectors so that it rings a bell when they register at the same time. It's called a coincidence counter.

ARTHUR ZAJONC: But the bell has no meaning as a coincidence detector unless it's placed in this larger context. I could also have some monkey hitting the bell. [*After the laughter settled down, I continued.*] A couple of things are going on here. One is the question of giving a meaning to a particular experimental result. This, you could say, requires a conscious mind. There can be automatic registrations, but the meaning is somehow connected to a person seeing something, making a cognitive judgment.

Let me emphasize the question of whether or not these quantum correlations are purely subjective. One of the things that tends to make us feel it is not subjective is, for example, if I can use this fact about nature to run a machine. If so, then somehow we feel that quantum entanglement works in the world objectively. This is not a question of cognition and philosophy but simply a question. Can it, namely, quantum entanglement, be harnessed to do any work? And it can. A new kind of quantum computer has been proposed that depends on this bizarre registration of coincidence events. It promises to be a very powerful technology and it depends, by its very nature, on nonlocality and entanglement. To my mind, this marks a watershed. All of the machines of the present day are classical machines, built like clocks—like the ones you used to take apart as a child. But this new kind of machine works on the principles of the new physics. Having a device that operates on the principles of nonlocality and entanglement proves that it is not just a subjective way of thinking about the world but that the world really is structured in this way.

GEORGE GREENSTEIN: You asked earlier about the distinction between subjective randomness and objective randomness. They are absolutely different, and we need to appreciate that. Subjective randomness is ignorance. Because we don't know everything, we don't realize the pattern and can't predict exactly what's going to happen. In the objective randomness of these experiments, there is no reason why these things are happening. There is no possible explanation, no reason at all. Anton, maybe you could explain the evidence for objective randomness.

ANTON ZEILINGER: In the case of these two correlated particles, when we try to make a model that explains why each particle does what it does, the model gives a different prediction for correlations than what we observe if the polarizers are not parallel. This is the strongest argument we have, and for three particles it gets even stronger. I should warn that there is one other way to "understand" the world without any randomness, and that is to assume that the whole universe is completely deterministic. If the whole universe, including my actions as an experimenter, is completely deterministic, then the problem doesn't come up. The problem does not come up because my actions in choosing a certain parameter are predetermined. The source knows beforehand, the photons know beforehand what will be measured, and so on.

The Question of Determinism

DALAI LAMA: Are there some philosophical schools in the West that do simply accept universal determinism?

ANTON ZEILINGER: I know one physicist who claimed that he could explain everything deterministically. I told him that was complete nonsense, rubbish, and he was upset. He asked why I had insulted him. I said that I hadn't insulted him; I had just said what was determined in that moment.

This brought a laugh from us all, but I wanted to contrast the nineteenth century's stand on determinism with the more nuanced view taken today.

ARTHUR ZAJONC: On a more serious note, while it was common in the past to hold to a complete determinism, with the advent of quantum mechanics and relativity theory I'd say this position is held much more seldom. You don't find very many people who hold to a position that is completely deterministic.

DALAI LAMA: When speaking of determinism, can you have a limited determinism? Can you have a deterministic view in regards to a specific and limited sequence of events? If this happens, then this will definitely happen. Or when you speak of determinism, are you necessarily inferring that the entire universe is locked into step?

ARTHUR ZAJONC: Quantum mechanics is deterministic in a slightly different sense, at an abstract level. Nowadays, when one says that quantum mechanics is deterministic, it means that there is a mathematical function whose evolution over time is completely lawful and deterministic. If you believe that this mathematical function is the description of reality, which some people do believe, then you would say reality is deterministically evolving. The problem arises when you come down to the real world—when you wish to make a measurement. When you perform a real experiment, that's where the randomness comes in. But Schroedinger and a number of other great scientists would say that at the mathematical level it is completely causal and deterministic.

DALAI LAMA: Is this because that kind of language, that mathematical description, has to obey the fundamental laws of logic?

Once again the Dalai Lama has identified several key issues in the foundations of physics. First, in what sense can there be a limited or partial determinism? Second, if mathematics is a completely logical and precise language of description, then must it not of necessity be deterministic? And behind these questions stand the two tenets of Buddhist philosophy—that no event happens without a cause and that no object, be it table or photon, has ultimate existence. Let us take these issues in turn.

By the early nineteenth century a fully deterministic picture of the universe had gained widespread currency in Western scientific circles. In 1812 the French mathematician and scientist Laplace made his famous declaration that if a Divine Calculator could know the velocities and positions of all the particles in the universe at a single instant, then he could calculate all that had happened in the past and all that would happen in the future. From initial conditions, the Divine Calculator could determine the positions and velocities of all particles at any point in time. Of course, the required knowledge of initial conditions was (and remains today) a practical impossibility. Likewise, even given this information, the performance of the calculation by any human intelligence is inconceivable, but this does not mitigate the force of the assertion as a principle. It should be mentioned that deterministic systems can show behavior that is remarkably chaotic. This is the basis for the modern study of deterministic chaos, which is to be strictly distinguished from true randomness of the type encountered in quantum mechanics.

Laplace's deterministic view persisted until the early decades of the twentieth century, when doubts were expressed about the possibility of such a calculation, even in principle. The cause of these doubts was Heisenberg's uncertainty principle, which states that one can never simultaneously determine the velocity and position of a particle. Finally, the uncertainty principle raises a still deeper question. Having measured the velocity, we say that the position of the particle in question becomes indeterminate; but what does this mean? Is the position merely unknowable by us, or does it in some sense cease to exist? Classically, we conceive of objects as existing at a place. What if the measurement of velocity destroys any meaning in the notion of a place where the particle is located? Moreover, this problem is not confined to position and velocity but is true for all pairs of complementary observables. Does the uncertainty principle then undermine our traditional notion of objects as having enduring properties and deterministic behavior?

Einstein originally had suggested the experiments with correlated pairs of entangled photons as a means of getting around the problems posed by the uncertainty principle. Ironically, the results of EPR experiments have only underscored the failure of traditional notions of particle identity. All this lies behind the Dalai Lama's statement that he sees some parallels between quantum mechanics and the Madhyamika philosophy of Nagarjuna, which also questions the independent, objective, and unconditioned existence of objects.

The question remains, To what extent can determinism still be a valid description of the world? Quantum mechanics gives a subtle reply. The mathematical description provided by Schroedinger's equation, for example, is entirely deterministic. However, the terms that appear in that equation, in particular the "wave function," do not correspond directly to any observable feature of nature. Therefore, a second step is required to bring *the mathematical theory into relationship with what we see. This step is* acausal; *that is, it breaks the determinism of the mathematics, which the Dalai Lama correctly identifies as a property of mathematics generally. This acausal feature of quantum mechanics is variously called the collapse of the wave function or the measurement problem. Many strategies have been advanced to avoid the collapse, but quantum mechanics has resisted all such resolutions. Physicists have learned to live with the hybrid nature of quantum theory: partly deterministic and partly nondeterministic. Theory gives the general behavior of events taken statistically but fails to determine individual measurement outcomes on individual particles. With this background, we rejoin the conversation.*

ARTHUR ZAJONC: The fundamental laws of quantum mechanics can be embodied in specific mathematical equations—for example, in Schroedinger's equation. These are just like other mathematical equations in physics, but the function that Schroedinger's equation governs is not a function that immediately describes the tangible realities of this world. It's the so-called state vector or wave function.

DALAI LAMA: If determinism only applies to this abstract level, but it doesn't have any real bearing on your experiment, then it's meaningless.

ARTHUR ZAJONC: It has a bearing on the experiment. Statistically it gives you the general form to expect, but it doesn't determine the individual events.

DALAI LAMA: This is fascinating. Buddhism has just this same problem, the same headache. Imagine an individual is in a situation where he can choose between different moral actions. If he chooses one, certain karmic consequences would follow. If he chooses another, then different consequences would follow. The Buddha would know these possibilities, but what actually wound up being the case would depend on what causes and conditions contributed to the situation. The Buddha would see the possibilities, but you would have to wait and see what actually happened.

THUBTEN JINPA: But one of the epitaphs of the Buddha is the "simultaneous knower of the three times." If you simultaneously are the knower of past, present, and future, you are not only at this point of possibility but also you see that later this happened in reality, and therefore this happened earlier.

ALAN WALLACE: It's an issue for the Christians, as well as the Buddhists, as soon as you have the notion of omniscience.

DALAI LAMA: There may be special anomalous circumstances relating to the omniscience of the Buddha. We should not take that as an absolute.

Taking Quantum Effects Personally

In the last phase of our conversation, the Dalai Lama returned to my opening remarks for the afternoon. I had expressed my personal sense of the profound and continuing impact that quantum phenomena have exerted on the thinking of physicists up to and including our own day. In other words, I take the results of quantum physics to heart. This was the entry point for an entirely different kind of discussion about the ways in which knowledge can change us personally.

DALAI LAMA: You mentioned, Arthur, the astonishment in reaction to these quantum mechanical experiments when they first arose, and you and Anton have both pointed out that you still feel this sense of astonishment. I feel that in the last century physics has been seeking out the nature of the phenomena in question from their own side—what really is the nature of light, for example—and then not finding it. And that is astonishing, not to find what you thought you would find, if you looked carefully enough. This astonishment pertains to photons, to electrons. But what impact has this astonishment and the implications of this not-finding had on your lives and attitudes or the lives of other physicists?

ANTON ZEILINGER: People have tried to make models, but this has not led to a new development yet. We physicists are immodest. We are impatient to find something new. I'm personally convinced that there is something new behind quantum mechanics in the following sense: that even as the world is strange, and I am personally convinced that it is strange, I want to know why it is so strange. There could be a reason for it. I feel that the only hope of finding it is to free ourselves from any concepts which are not absolutely necessary. Use the minimum that you can. What are the concepts we are really talking about? The concepts are just these pieces of stuff—the apparatus—and the click of the detector. These are the things we can really talk about. The models we develop in quantum mechanics, the explanations with equations and so on, are just a way to write down what we know about the apparatus now and what we predict about future features of the apparatus, namely, the click. Then the next step comes in, which is the question of what it means in the broader sense. To me, it means that this apparatus behaves in a strange way. It clicks at a certain time without any reason why it should click at that time. It might equally have clicked one second before and one second after. This is strange.

DALAI LAMA: I'm convinced that there have to be some hidden variables there. I won't attribute it to karma. There has to be some kind of

purely physical contributing factors there that have not been identified yet. They have to come from somewhere, from outer space for all I
know.

Could we come back to my question? How do these astonishing
things affect you? You are still talking about this stuff, this apparatus,
which you can put in a box and forget about. But when you put your instruments away, do these ideas affect the way you view your daily life,
the larger world?

ANTON ZEILINGER: Oh, yes.

DALAI LAMA: How? The scientific analysis of the nature of reality leads us
to a point where even the notion of reality tends to disappear. So we are
in fact almost compelled to refer to objects as this so-called table or this
so-called microphone. Does that kind of awareness have some impact,
for example, on how you look at a beautiful flower? Normally, if we see
it as something absolute, beautiful from its own side, we would feel
more attachment to it. On the other hand, once we see that it is not substantial, not absolute, we would see it with less attachment: "Oh, this is
a so-called beautiful flower." In Buddhism, this view of emptiness has a
profound ethical impact on the way a person would live his or her spiritual life. Emptiness has a direct relevance to the person's worldview and
relationship to the world.

ANTON ZEILINGER: Maybe I can answer personally. Looking back at the
physics of the last century, the idea was that it was the century of mechanics. The idea was that the world, including us, is just a big mechanical machine that evolves according to certain laws. I find this a
very boring view, a very sad view. I find the new view much richer. The
world is much more open because there are things happening that have
no cause. You cannot explain them in a mechanistic way. So how do we
now view the flower, the microphone, or whatever? My point of view
(which I hope agrees with Bohr's, but you never know) is that these
things, these everyday life experiences, were here before physics. They
are evidently here, we take note of them, we use them, and we speak of
them. We have to use them in our language because we have to communicate. They come before physics.

DALAI LAMA: This is quite true. Even when Buddhists talk about emptiness, the very fact that the word *emptiness* is used indicates that we are
talking about the empty, nonsubstantial nature of something that exists.
Otherwise it makes no sense to talk about emptiness.

*In Buddhism, when inexplicable situations involving humans appear, one
usually inquires about karmic causes. With electrons and photons, karmic*

causes seemed implausible to His Holiness. But the real question he posed was, What effect has quantum physics had on our lives? Anton's answer had two parts. First, quantum physics experiments have prodded him to challenge every one of his assumptions about the nature of reality. Second, he works to go deeper, striving to see what stands behind the strangeness of quantum phenomena. In this he reasserts in his own fashion the hope spoken of by Weiming at the outset of the afternoon, the hope of seeking "a totally different view of the world." Of importance for the Dalai Lama in all this was the modern scientific evidence pointing to the absence of an absolute reality of fundamental entities with enduring properties. As he said, if we grant that beauty is not absolute, not given from its own side only, but rather dependently arising, then this insight can lead to the wisdom of nonattachment and thus liberation from suffering. Although Anton was not ready to follow the Dalai Lama to his final conclusion, he did reaffirm the priority of the given world. The world of experience is not to be explained away by deterministic mechanics or its modern successors. Before physics, the world of cloud and rainbow, of children and songs, existed. Beauty came before physics, reality before emptiness.

3

Space, Time, and the Quantum

*On the morning of the second day, David Finkelstein continued the intro-
duction to quantum mechanics, but now with a change of emphasis. He
began by building on the foundations laid by Anton Zeilinger on the first
day, but by the late morning he was describing the new ideas of space and
time that both quantum mechanics and Einstein's relativity theory demand. The Dalai Lama had worked with David before, and I had noted
then, as now, the special delight they seemed to take in each other's company. Even at those places in David's presentation that became difficult for
all of us, the Dalai Lama always attended to him with interest. Not infre-
quently David would savor a small irony that arose in the course of his
presentation, his eyes sparkling with the humor of the situation. These moments lifted the session, which was intellectually the most demanding of
the entire week. But we had come to work hard, and on Tuesday, under
David's tutelage, that is exactly what we did.*

*Like Anton, David is uncompromising in his intellectual integrity. If we
are to take physics seriously, then we must discover its implications for our
thinking about the world, and once the implications are known, then we
should apply them consistently. If quantum mechanics demonstrates that
our traditional style of thinking is inadequate or even mistaken, then we
must change our thinking. Far from treating quantum mechanics as a
purely calculational tool, David sought to present the implications of the
new physics for the very way we should think and speak about the world.*

In this regard, quantum physics not only contains confusing puzzles but also gives positive guidance.

In addition, there is the issue of preexisting and enduring properties of quantum systems. Briefly stated, it seems that at least microscopic systems do not have these properties. Instead of thinking of the objects of the world as having preexisting properties that we discover by passive observation, David suggests a more active picture, one in which the action or operation of observation is essential to the determination of the property. The shift is dramatic. In place of objects or states, the emphasis is now on action and transformation. The latter are primary, the former derivative, David argues. What are the implications of this action orientation for our thinking about the world? What is the new logic that obtains? And what is the new view we should have of space and time?

As a theoretical physicist of the first order, David uses language of description that is occasionally abstract and even mathematical. Through my notes and footnotes I will attempt to fill in the background needed to follow the presentation even if the reader is not mathematically inclined.

From Fixed Classical States to Unfolding Quantum Actions

DAVID FINKELSTEIN: Yesterday Anton showed us some confusing experiments related to quantum theory and helped us to feel the distress of physicists trying to understand such a paradoxical view of nature. It was pointed out that there must be a positive side to quantum theory as well, where it stops being distressing and starts giving pleasure.

I thought today that I would stress the positive side of quantum theory. We began our meeting with the principle of skepticism. In order to be skeptical of something, first you must notice it. It's very hard to be skeptical of things you do not notice. The hardest part of each dramatic change that has occurred in physics since 1600 has been to become aware of the assumptions of the old theory that had to be given up. The most confusing state of affairs is during the early days of a new theory, when you still cling to some of the old assumptions, and yet some rewards of the new theory are attracting you further. I used to think of 1924, the year in which Heisenberg discovered the quantum theory, as a kind of abyss, a Grand Canyon, separating the old physics from the new, or a desert separating two fertile regions. But this is too symmetric. The two sides of an abyss are on the same level. The two sides of a desert are symmetric in respect to each other. Really we should regard this as a change in level, an evolutionary step: Quantum theory is on a higher plateau than the older physics. Those on the lower plateau find the upper one invisible, mysterious, confusing. When you reach the upper plateau, you see the lower one and recognize it as part of a much larger picture. Usually, when you hear of a paradox or an inconsistency in quantum theory or any negative statement about it, it is from the point of view of the old theory. It's true that quantum theory has its own problems—unfortunately we do not have a perfect theory yet. We are still working toward the next theory, but I'll try to say a little about the direction where I think the next plateau lies.

Let me begin by emphasizing the main points of classical physics that have to be given up. The central idea of classical physics is the idea of the state. A physical system is supposed to have a state. In mechanics, for example, the state of a planet is defined by exactly where it is and how fast it is going. The state is that information about the planet, about the system in general, which sums up everything about the past that is necessary for the future and tells everything about the future that can be determined by the past. It's a meeting of our acts of preparation and our acts of prediction. All the laws that we use to predict the future are based on the idea of state and have experimental consequences.

We throw a ball; someone else succeeds in catching it. To account for this, we introduce the auxiliary idea of the state and we follow the state through the entire process. The state is an auxiliary concept introduced to account for experiments that work every time. From the idea of the state, we build the idea of a predicate or property of the system. We represent the system as a set of states—those states in which it has a given property, as opposed to those in which it does not have the property. This representation of predicates implies many of the laws of classical logic. The idea of "and" comes from the intersection of two sets of states. The idea of "or" comes from the union of two sets of states. The idea of implication comes from the inclusion of one set of states in another set of states.

The idea of the state is also used to describe actions upon the system. The most elemental action is defined by an initial state and a final state. More general actions can be defined by a table showing for each initial state the frequency with which a particular final state occurs. Such a table of transition frequencies is also called a matrix. Its elements are essentially probabilities. In cases where we know the most, the elements would be zero or one: zero, meaning the transition from that initial state to that final state never occurs; one, meaning it always occurs. Such a table would be appropriate, for example, for experiments with a coin.

For the purposes of the experiment, the coin would have two states: heads and tails. The things you do to it might cause it to change from heads to tails. This change would be represented by a matrix with a single one where the heads column meets the tails row.

Here and later on David uses concepts drawn from mathematical physics that may not be familiar to some readers. Simply put, in this context a matrix is an array of numbers that mathematically describes how a system is transformed from one state to another. David uses the example of a coin, which, of course, has only two possible states: It can be either heads or tails. The matrix connecting the initial to the final state has four elements (2 × 2 = 4). The state heads can be represented by a column matrix (also called a column vector), with a 1 in the top position and a 0 in the lower position. The tails vector is like the heads but with the entries exchanged. The operation or action that transforms heads to tails is given by a third square matrix. I give all three below.

$$\begin{matrix} \textit{Heads} & \textit{Tails} & \textit{Transformation matrix} \\ \begin{pmatrix} 1 \\ 0 \end{pmatrix} & \begin{pmatrix} 0 \\ 1 \end{pmatrix} & \begin{pmatrix} 0 & 0 \\ 1 & 0 \end{pmatrix} \end{matrix}$$

Those readers who know matrix multiplication can see that operating on the heads vector with the transformation matrix yields the tails vector.

This idea, that a transformation can be described mathematically by a matrix, is quite general. For example, the system under study could have been a box instead of a coin. The box can be rotated mathematically any way we wish by operating on it with a matrix. Finally, in quantum mechanics, too, a matrix can effect the transformation of one quantum state to another.

DAVID FINKELSTEIN: The great logician George Boole, who invented the Boolean logic and Boolean algebra used in computers today, began his work by trying to define exactly what a class—or a property or a predicate—is. With each class, he associated an act of "election," as he called it, or choice. A class is defined by the act of selecting from a general population those members of the population that have a given property. Thus, built into the classical idea of the state is the consequence that predicates (or filters for predicates) change no property of the system being tested. They simply pass or reject: What gets through has the same properties as what entered. You can think of this as a passive principle. Logic does not act; it simply selects.

Associated with the idea of a state is a special kind of relativity. If things have exact states, then all that differentiates two observers of a thing are the names that they associate with the states. Each word used by one experimenter corresponds to a unique word used by another experimenter. I say *here*, you say *there*, but we mean the same thing.

The first indication that this idea of state does not adequately describe reality begins with quantum theory. The great leap to quantum theory happened in 1924, but all great theories spend a long time being born. Once quantum theory was invented, we could look back and see all the attempts to approach it before. I've been able to trace this as far back as Aristotle. The logic that today we call Aristotelian is not the logic that interested Aristotle. It is good only for static things. He spent most of his effort trying to understand a special logic appropriate to things that are changing. He felt that when something changes from having a property to not having a property, there is a moment in between when, in some sense, it both has the property and does not have the property. According to Aristotle, when an egg is changing into a chicken, there is a moment when it is both egg and chicken and neither egg nor chicken.

The Phenomena of Wave Interference and Polarization

DAVID FINKELSTEIN: The first experimental indication of a crack in the ancient structure occurred in the age of Newton. When Newton tried to explain the whole world in terms of objects with states, he was aware that light sometimes behaves like a particle and sometimes behaves like a wave. In fact, it's not clear what experience he had in mind when he insisted that light is a particle. The evidence at the time was not very clear, although in hindsight we can see that he should have been able to make that deduction because vision itself is a form of the photoelectric effect. The same effect that Einstein used to discover the particle nature of light was already waiting for Newton.

But in fact, he did have direct experimental evidence of the wave nature of light. Newton's rings are made when two pieces of glass come very close to each other. You see the little rings. . . .

The photoelectric effect refers to the phenomenon of electron emission from surfaces when they are illuminated by light. Light falling on the eye stimulates an analogous process in the rods and cones or the retina. Einstein received the Nobel Prize for his theoretical analysis of the photoelec-

tric effect. David showed the Dalai Lama two pieces of glass held close together but with a very small air gap between them. One plate is slightly curved, so the spacing gradually changes between the two glass plates. Light is partially reflected from the air-glass interfaces and makes its way back into our eyes. A set of delicately colored, concentric rings can be seen in the glass.

DALAI LAMA: Did Newton actually invent this?

DAVID FINKELSTEIN: Since they are called Newton's rings, you can be sure he didn't invent them. Apparently they were discovered by Hooke.[1] In the center of these rings, the two pieces of glass are in contact. There is some light reflected there from the top piece and some reflected from the bottom, and the two are out of phase. They cancel each other, to some extent, making a dark spot. At the first bright ring, instead of being out of phase, they are now in phase. The two pieces of glass have moved apart just enough to fit a whole wavelength of light between them. That is about one-millionth of a meter. This is so small that just by pressing the glass, you bend it enough to move the rings around. You can measure distances in units of the wavelength of light with two pieces of glass.

The other important experimental indication that Newton had for the wave nature of light dates back to the Vikings. At about the same time that the Chinese were learning to navigate with magnets, the Vikings were learning to navigate with polarizers.[2] In the North Sea, it is often hard to see the sun. That makes it hard to find your way home after you have raided Normandy. So, at such difficult times, the captain, or suitably bearded old man, would hold a crystal up to the sky and turn it and tell the direction of the sun even though he couldn't see it. It is a little easier with modern polarizers: If you look at the sky through this polarizer and turn it, you will see that the sky is brighter in one position. There is a slight change in intensity in the light from the sky.

DALAI LAMA: There is something wrong with my eyes; I do not see it.

DAVID FINKELSTEIN: It is a very small effect. You have to turn it rather slowly: It happens in a small interval. If you turn it quickly, you go right through it.

DALAI LAMA: Ah, yes, a very slight effect. Why is there no difference in the coloration of the polarizer?

DAVID FINKELSTEIN: Yesterday we talked about color blindness and what it would be like to have colors that we cannot see. This is such a blindness. Our eyes are not sensitive to polarization, but by using this device we can at least become aware of it. Other animals, like the bee, are very sensitive to polarization, but it has not been important for people.

To account for the polarization of the photons that come from the sun, Newton said they have coasts, or sides. If you want to think of a photon as an object, do not think of it as a point. Do not think of it even as a needle, but rather as an arrow. An arrow has fletches, or feathers, and their alignment differs from one to the next. A polarizer will transmit photons with a particular alignment and block other photons.

DALAI LAMA: I am trying to understand this discussion of photons and polarization in the light of what Anton was discussing yesterday. Yesterday he described the entanglement of two photons being emitted simultaneously from a single source. Is the sun not one source? Does that mean that all of the photons being emitted from this rather large source are also entangled? Do the photons emitted from the sun have a random polarization, or are they all connected in an orderly way?

DAVID FINKELSTEIN: The sun has to be regarded as a great many sources. Each atom on the sun emits independently, so there is a very confused polarization. A laser is a coherent source. The sun is a highly incoherent source. The actual polarizers the Vikings used were crystals like these of Iceland spar, and here I've made a little pinhole on one side. There's only one hole, but if you look at the other side through the light, you will see two holes. If you check, you will find that each of the two has a different polarization. If you turn the crystal or the polarizer slowly, first one hole will disappear and then the other. Newton resisted the great temptation to account for this with the wave theory and insisted that each photon had two states, vertical and horizontally polarized, let us say.

The crystal Iceland spar, also known as calcite, has the property of double refraction. This means that when an object is viewed through the crystal, its image is doubled. Instead of one pinhole, for example, two appear. In addition, the polarization of the two images is different. David demonstrated all this to the Dalai Lama. He then demonstrated Malus's law concerning the intensity of light passing through two polarizers. If the two polarizers are parallel, then all the light that gets through the first also gets through the second polarizer. If the two are perpendicular, then no light gets through. Malus discovered the exact empirical law that tells us how much light will get through for any angular orientation of the two polarizers, without knowing anything about what exactly causes this phenomenon.

The Failure of Microscopic Causality

DAVID FINKELSTEIN: Then 100 years later, the engineer Malus considered the question of how many photons that get through one polarizer will succeed in getting through another one at a different angle. (It's not clear whether he did the real experiment or a thought experiment; in those days, there was not the rigorous distinction between the two that we make today.) If the two polarizers are parallel, ideally all the photons that get through the first also get through the second polarizer. If they are perpendicular, you can see that virtually none of the photons get through the second polarizer.

DALAI LAMA: Are there not many gradations of polarization in between the vertical and the horizontal?

DAVID FINKELSTEIN: In fact, every orientation of the polarizer defines another kind of polarization. One has here a paradox that seems to be inconsistent with the ordinary idea of state. For each experiment, the photon behaves as if it has two states. The photon either goes through or it does not go through. It never splits, and yet, in some sense, you determine what those two states are by how you hold the polarizer. It is a peculiar blend of two discrete states and the continuous. This is the typical situation of quantum theory.

In about 1800, Malus broke with Newton on the question of how to interpret these photons going through two polarizers. Newton insisted on following in detail what the photon did. He wanted to know how the photon made up its mind, having gotten through the first polarizer, whether or not to go through the second polarizer. In order to account for the action of the photon, Newton invented mechanisms that involve a guide wave or pilot wave accompanying the photon to account for Newton's rings, for example, and for the fact that photons don't all behave the same way when they reach the polarizer. Malus wiped out all that thinking. He simply described the experiment. When one polarizer is at a particular angle and a second polarizer is at a different angle, such-and-such a fraction of the photons get through. All study of motion in quantum theory today takes this form. There was a lapse of over a century between Malus and Heisenberg, but essentially we have returned to Malus's law. We only speak in terms of an entire experiment. We do not ask questions about what is the state of the system. We describe how you begin the experiment, how you end the experiment. We may interpolate many, many intermediate stages if necessary. But we don't look for the mechanism. We simply tell the probability of the photon getting through the system. Sometimes we are sure the photon will not get through; sometimes we are sure it will. Most of the time—almost always—we are in neither situation. Almost always, quantum theory does not predict the result of the individual experiment.

Here David is emphasizing the failure of quantum physics to describe the details of a microscopic causal mechanism that produces individual quantum events. Quantum theory can predict statistical features and give a causal account at that level, but it fails to give what Newton and Laplace saw as the goal of science, namely, knowledge of the positions and velocities (i.e., the state) of the particles under study. Instead physics retreats to the more modest position of predicting the outcomes of experiments in a way analogous to that of Malus in his study of light passing through polarizers.

The next point David wishes to make is subtle but central to his argument. Quite naturally we all think in terms of enduring properties of the objects around us. The glass in front of me is made of sodium silicate; it rests on the table just to my right; and so on. It doesn't matter whether I am looking at the glass, measuring its properties. These properties belong to the glass and are quite independent of my observing activity and me. But in the world defined by quantum physics, these sensible assumptions no longer hold. The underlying state of a system (my glass) really presupposes a set of measurements that determine the glass's attributes. These measurements are operations that can be represented by matrices, as we have seen before. David is emphasizing to the Dalai Lama the importance of the shift away from conceiving of the world in terms of underlying states or properties, and he suggests that instead we think of the primary underlying reality as action, as operations. In addition, the particular kinds of operations that are characteristic of quantum physics are different from those given by classical physics, and therefore the matrices used to describe these operations mathematically are themselves different, as Heisenberg discovered. Associated with these differing sets of operations are differing logics or ways of thinking about the world.

The Logic of Operations

DAVID FINKELSTEIN: When we think in terms of underlying states, we commit ourselves to a certain concept of operation. It's possible to see the states, so to speak, in the pattern of the operations. The pattern of operations that we find in nature is different. It is not the kind of pattern that arises from underlying states. So, we no longer talk about the states; we talk about the operations.

Heisenberg began his work in 1924 by saying what the form of a general operation was. We saw that even in classical physics you could represent an operation by a table, a matrix. But in classical physics, the numbers in the table all have to be positive because they represent probabilities or frequencies. Heisenberg simply let them be positive and negative. It's still a puzzling question, why exactly this concept of operation always works. It's really the basic assumption of quantum theory. (Because matrices figured so importantly in quantum theory, at first it was called matrix mechanics.) Many people puzzle over this. I won't go into it today. I'll simply start from that point.[3]

Once you have a complete theory of operations, you can deduce a logic from it. Every predicate, every class, is associated with a filter operation. For example, the first polarizer singles out a certain class of photons: those which get through. A differently oriented polarizer singles out a different class of photons. From two such things, you can form an idea of an "and" combination and an "or" combination. The "and" of two classes is represented by yet another filter, which transmits only photons that are sure to go through whichever of the other two filters you choose to apply. "A and B" is a filter that produces photons which are sure to have the property A and sure to have the property B. "A or B" is a filter which is sure to be passed by photons having property A and also sure to be passed by photons having property B. Similarly, you can define negation. Then you can check the laws of Aristotle experimentally. You find that all of Aristotle's laws, including the law of noncontradiction, hold. For example, either A or not-A is true. Both A and not-A cannot simultaneously be true. This law holds. However there is a simple law of elementary logic that is violated.

In logic, when one negates A, one gets not-A. In terms of polarizers, if A corresponds to vertical polarization, then not-A is horizontal polarization. Likewise, the orientation of the not-B polarizer is perpendicular to the B orientation. Noncontradiction, or Aristotle's law of the excluded middle, states that either A must be true or not-A must be true. David is not suggesting here that the law of the excluded middle is itself violated but rather that a certain application of Aristotelian logic more generally fails in quantum mechanics. The standing of the law of the excluded middle in quantum mechanics will be taken up more fully later on.

DALAI LAMA: Is this logic being presented as a universally applicable principle?

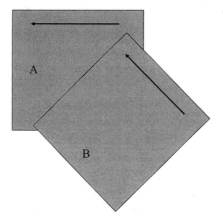

Figure 3.1 Polarizers A and B are in tandem and are oblique to one another. Light enters polarizer A from the rear. The light that exits polarizer A is horizontally polarized. The light that exits polarizer B is half as intense and is polarized at an oblique angle (45 degrees). The orientation of the polarization of light is determined by the final polarizer.

DAVID FINKELSTEIN: I don't believe in universal principles. It works today.

DALAI LAMA: But does this logic apply only for a specific system or situation, or is there a general applicability for it?

DAVID FINKELSTEIN: The quantum theory is supposed to be applicable to any system. In that sense, these are general laws.

Let me return to the simplest law of logic that breaks down in quantum theory. If you have two properties, A and B, then A or not-A is true, and B or not-B is true. Classically you would say that one of the following four cases must hold: "A and B" or "A and not-B" or "not-A and B" or "not-A and not-B." It's very easy to demonstrate with polarizers a situation where all four of these are false: "A and B" is false; "A and not-B" is false; "not-A and B" is false; "not-A and not-B" is false.

David then held up two polarizers together, with one turned to an oblique 45-degree angle of orientation relative to the other (see figure 3.1).

Let me exhibit it here. Let this polarizer be the filter for A, and let this one, at a 45-degree angle to A, be the filter for B. "A and B" would be a filter, all of whose output is guaranteed to pass through both polarizers

A and B. But there is no such thing. If they all get through A, then Malus's law tells you half of them will be blocked by B.

DALAI LAMA: How would you identify a photon that has a property of both A and B polarizers? We began our discussion by identifying two different orientations of the photons that travel. If one is A and one is B, how can one be both A and B?

Complementary Properties Can't Simultaneously Exist

DAVID FINKELSTEIN: We're looking at the case where A and B correspond to oblique polarizers. If you keep the polarizers parallel or perpendicular, the classical logic works. The oblique case corresponds to just what Bohr called complementary properties, and in this case quantum logic differs from classical logic. Here "both A and B" is supposed to be a property of a photon. It is, therefore, defined by yet another filter. It is a hypothetical filter, of which all of the output will surely pass through an A filter and will also surely pass through a B filter. And there is no such thing. If someday we make one, then we will change our logic.

I then asked David to give a classical example in which an object possesses both property A and property B and, therefore, always gets through the two filters in tandem.

DAVID FINKELSTEIN: If we are looking for red Macintosh apples, we can imagine an assembly line that the apples go down. Some of them are red and some are green. Some of them are Macintoshes and some are Baldwins. A person first selects the Macintosh apples and then selects the red apples. Those apples will surely pass the Macintosh test, and they will surely pass the red test. Those are not complementary variables. If you try this with photons, it doesn't work.

DALAI LAMA: Can you explain why the other three options are invalid?

DAVID FINKELSTEIN: The point is that all four of the terms I've mentioned are oblique in the same way. A and B are at 45 degrees; A and not-B are at 45 degrees; not-A and B are at 45 degrees; not-A and not-B are at 45 degrees. They are all oblique, and in the oblique case, the conjunction is false. Usually it is meaningless to say a photon is polarized this way *and* polarized that way. But it's also accurate to say that it is meaningful but false. The experimental evidence shows it is simply false to say that the photon is polarized this way and also polarized that way. And that means the conjunction is false.

In the above discussion, David demonstrates that we cannot select photons with particular polarization states in the same way that we can select apples with particular properties. Classical measurement is a simple sorting according to preexisting attributes. By contrast, the attributes of a quantum system cannot, in general, be thought of as preexisting. Rather, the act of measurement acts on the photon to call forth a polarization. Subsequent acts of measurement do not simply select for another polarization independently; rather the final measurement calls forth polarization along the new axis in accordance with Malus's law.

With this established, we paused for a lively tea break. In fact the conversation seemed only to intensify during the break. As the audience wandered off, a small group huddled around the Dalai Lama to debate the intricacies of Western and Buddhist logic and the role of empirical evidence. In our common world of experience, and in the classical science that is based on that experience, we do not encounter inconsistencies. Aristotle's logic works. But with quantum phenomena, we encounter situations in which empirical evidence confounds our simple logical assumptions.

We returned to the issue of how we know something is true.

We Change What We Interrogate

DAVID FINKELSTEIN: You might know because something is theoretically required. You could speak of something being not merely true but necessarily true. Or you might know simply by observation that something might be possibly true. We can know a room is white, even though it is dark, because we painted it white in the past. Or we can know a room is white because we look at it, after the fact. We can know it's white by an initial act (recollection of painting it white), and we can know by a final act (looking). In classical physics, these two are always consistent. In quantum physics, we have new types of logic that were never considered by Aristotle. In general, when we consider analogous situations in quantum physics, the initial and final acts are inconsistent.

I used to think this was very fundamental because I learned logic before I learned the rest of physics and mathematics. Whitehead and Russell said that logic was the basis of all mathematics and, therefore, all physics. I no longer believe that. I think looking at the operations is more important than looking for the true properties of a system. This is partly of historical interest for me. I went through this many years ago. The important thing is that one can no longer imagine that asking questions does not change what you interrogate. A photon is going to be changed by going through a polarizer. All the filters of quantum mechanics have that characteristic. All the paradoxes ultimately come from the fact that asking a question changes the system.

DALAI LAMA: All of the notions of logic, such as conjunction and disjunction, are based on the idea of class and types, many of which are actually, in some sense, mental constructs. They are tools to understand the relationships and the functions of the material world. Does the logic you describe really have any applicability in the realm that you are talking about?

DAVID FINKELSTEIN: That's an important point. The fact is that Heisenberg's laws or Malus's law are the ones we use. The logical laws have not been applied to scientific problems very often.

DALAI LAMA: When you speak of conjunction within Buddhist logic, it is basically straightforward set theory. Set theory is used a great deal in Buddhist logic. For example, one set may be included within another. Take the question of whether a photon is an expression of energy. The answer is yes. Are all expressions of energy necessarily photons? I believe the answer is no. It's called a three-point relationship: Anything that is a photon is energy, but anything that is energy is not necessarily a photon. But you're speaking of photons as general types of phenomena. You're not taking one specific photon and comparing it to another specific photon. You are dealing with another level, one removed. You are not dealing with this photon and that photon but photons in the abstract, in general.

DAVID FINKELSTEIN: Absolutely. The point is we can't use set theory because photons don't have states.

DALAI LAMA: Even though you don't speak of them having the same state because we're not talking in that kind of logic anymore, you do speak of them having the same operation, yes?

DAVID FINKELSTEIN: We assume that one can repeat an operation. But it is important that a photon does not have an operation.

[*Someone then interjected*]: Does a photon have a property?

DAVID FINKELSTEIN: There is no way you can tell from a photon how it was made. You can tell what its spin is. You can tell its energy. You can tell its momentum. But you cannot tell its source; you cannot tell how it was made. It doesn't carry an operation with it.

This again is related to the fact that we're operating on a different level now. And we are looking at assemblages. When we speak of an individual photon, we refer to the population from which it came. We're really talking more on the mental level than usual because we don't have access to the individual as fully as we used to think we did.

Reality or Actuality?

DALAI LAMA: At a previous meeting at Columbia University, one physicist made the point that in quantum physics the whole notion of reality becomes problematic and suggested that people should speak in terms of actuality instead. What is the basis for the new view of reality you present?

DAVID FINKELSTEIN: For me reality refers to an object which has states. From the viewpoint of quantum mechanics, we must not talk about the state but about the act. Reality is a succession of states. Actuality is a succession of acts.

DALAI LAMA: Yes. There is a very familiar parallel in Buddhist philosophy. For an individual — a self or person—if you seek out what is the state or sequence of states of the person, it's not to be found. But you can speak meaningfully of the functions, the deeds of the person.

With these remarks the tea break ended. Everyone came back into the room and settled in for the last part of the morning session, in which David applied the new quantum way of thinking to space-time.

Quantum Thinking and Space-Time

DAVID FINKELSTEIN: What I have been trying to do is just to get across the idea of a quantum way of thought, in contrast to the classical way. First we must have this way of thought before we can think about new things with it. My purpose then is to apply this quantum thinking to the structure of space-time.

Let me just mention that part of the transition from the classical to the quantum way is a change in the concept of relativity. Relativity is stretched in a way that is so far beyond Einstein's thought that he never was willing to accept it. In the relativity of type I, that I spoke of before, it's taken for granted that all observers see the same properties—that they all see the same state and give different names to it. Relativity before 1924 was type I, the relativity of Galileo, Newton, and the young Einstein.

In relativity of type II, this is no longer assumed. Each experimenter must make a choice of the properties into which he will analyze the world. One experimenter can use two polarizers such as "A or not-A." Another experimenter will use "B or not-B." That's all they can know. If either tries to know "A and B," it's false. If they try to know "A and not-B," or any of those special cases, it doesn't work. Each experimenter must make a choice. And a statement made by this experimenter simply cannot be translated into the language of that experimenter. Or in some cases, it can be translated but not uniquely. If I say that a photon is vertically polarized, then you may say that half the time it's polarized upward and half the time it's polarized downward.

So we have a new kind of dictionary, with a one-to-many translation, and each of the possibilities has a frequency associated with it. This is more like real life. In real life when you translate from one language to another, there are many possibilities. Some are common; some are rare; some are archaic. Little by little, physicists are getting back to humanity, to living in the world as it is rather than in the fantasy of Descartes.

David emphasized how each observer has a special viewpoint that is not uniquely translatable to that of others. What I measure with the instruments at my disposal cannot in general be expressed in terms of the outcomes you measure with your instruments. In Europe, temperature is measured in degrees centigrade; in the United States, we measure it in degrees Fahrenheit. Although the numerical values will be different for the two systems, there is a unique correspondence between them. In quantum mechanics, this is no longer the case. Your measurement does not correspond to one of mine.

If we also include the development of relativity by Einstein, then a still deeper and more fundamental feature of our world conception emerges. In what follows, David appeals to certain aesthetic aspects of the theories used by physicists, which are taken as an indication of the depth of the insight on which they are based.

DAVID FINKELSTEIN: Now let me begin to apply this concept of quantum thought to the structure of space and time. Again, I'll begin with the classical ideas of space and time as a starting point, to be conscious of the assumptions that we must painfully give up. Actually, sometimes giving up assumptions can be joyous. It's a question of where one is coming from and going to. I will look at classical space-time from the quantum point of view, so I will stress its negative features, but I emphasize that it has many, many positive features. It's a theory that works, and it's the best we have. Nevertheless, we should look at it for signs of the next theory.

The idea that time is something like a real number—exactly knowable and infinitely divisible—leads, in conjunction with the other principles of physics, to something like a theory of fields. We describe an event by two sets of quantities. First we tell where it is and when it happened. It's like giving the address of a meeting. But this does not tell yet who is going to be at the meeting. And so, after giving the location and the time, we give the values of the electric field, the magnetic field, the gravitational field—as many fields as we need to describe the state of affairs at that point of space-time. There are two sets of descriptives: space-time and field.

Until today, field theory has been the only way to combine the principles of locality and relativity. We satisfy locality by identifying points and writing laws which relate the field at one point only to those space-time points in the immediate vicinity. We then satisfy the principles of complementarity, that is, of quantum theory, by treating the fields as quantum entities. We no longer speak of their exact state but only of the operations that we carry out on them.

There are, however, various symptoms that field theory is not the final theory. One of them is degeneracy. There are several signs by which you can recognize a degenerate theory, that is to say, a theory which is only a projection of a deeper theory. For example, in the last century, people doing classical mechanics should have recognized that it was a degenerate theory. The main sign is that you have a physical entity which is reducible in that it contains another physical entity, but not fully reducible in that it also contains things that are not physical. Let me give an example of this complicated idea. In classical mechanics, a space-time point is a physical entity. Contained in it is the idea of a point of time, an instant of time, which is also a physical entity in classical physics. If you remove the idea of time from the space-time point, what is left over is a point of space. There is no such physical entity as a point of space. To put it another way, if you compare two observers and their descriptions of a space-time point, you see that time enters into the transformation properties of space, but space does not enter into the transformation properties of time.

May I give an example? I slept on the train from Delhi. (Apparently, I will never forget that trip.) In my frame of reference, the spatial distance I traveled between going to sleep and waking up is zero. For you to compute how far I traveled in your frame of reference, you must ask me how long I slept, and you must also know the speed of the train. So, time enters into space. But space doesn't enter into time. If I say I slept for four hours, you know instantly how long I slept. You also say I slept four hours. The fact that space does not enter into time is a sign of a degenerate theory. Along comes Einstein's theory of special relativity, where time still enters into the transformation of space, but now space also enters into the transformation of time.[4] The amount by which it enters is rather small for ordinary experience. If the train traveled near the speed of light, it would be quite substantial.

We have a similar situation in field theory today. Suppose I tell you exactly where I measure the field in my reference system, but I do not tell you what the field is. You can tell me exactly where I measured in your reference system. The field does not enter into the transformation law of space. But if I tell you that I measured an electric field of such-and-such a strength in this direction, you cannot compute your description of the electric field from that information without asking me where I measured the field. Space enters into the field transformation, although field does not enter into the space transformation. This is a sign of a degenerate theory.

One can easily see where this degeneracy comes from and how to lift it, just as one did with classical mechanics. If, instead of speaking of a point and a field at the point, I spoke of two points, then I would have an object which was fully reducible. It consists of this point and that point, which have separate laws of transformation. If you think of velocity, for example, as just an approximation of a discrete jump from one point to another and, instead of specifying a point and a velocity, identify two points, you lift this degeneracy.

Alan Wallace, who had studied physics, as well as Buddhist philosophy, interjected a question of clarification. It was one that quickly led us to a view of space-time that David has found to be more elegant and insightful. Along the way, David demonstrated the theoretical difficulties one encounters in trying to construct a consistent classical view of space-time. It simply does not work once we include a careful description of the measurement of space-time points. Having pointed out these troubles, David went on to advance his own view of a discrete space-time.

Of Discrete and Continuum Theories

ALAN WALLACE: Am I right in inferring that the very notion of a field is a vestige of state logic, as opposed to operations?

DAVID FINKELSTEIN: I hadn't said that, but you sense it perfectly correctly, yes. I was bringing up a slightly different point, which is that you can recognize a degenerate theory by the existence of entities that have asymmetric influences. A affects B, and B doesn't affect A. This suggests that the field theory we have is a degenerate version of a discrete theory, in which the discreteness is so fine that we have to approximate it by a continuum.

ALAN WALLACE: And a discrete theory pertains to points?

DAVID FINKELSTEIN: A point with finite separations, as opposed to points with directions. This suggests that underlying the continuum of space-time, there is something made up of discrete space-time units. In order to incorporate the continuous symmetry that space-time exhibits, we should imagine these units as quantum entities rather than classical.

Another disease of the classical picture is that it leads to infinite expressions for many physical quantities. This was originally the reason that quantum theory was introduced by Planck, in order to make the heat capacity of an oven come out to be a finite quantity. But there are still other quantities that come out infinite, entirely because of the assumption that there is an independent degree of freedom at every point of the continuum. There is simply too much going on in the continuum theory for a finite result to all physical questions. In the continuum theory, there seem to be an infinite number of variables in any region of space, no matter how small.

DALAI LAMA: What exactly do you mean by continuum theory?

DAVID FINKELSTEIN: The assumption that time is infinitely divisible, that it is represented by a real number. At each moment in time there are an infinite number of field variables, all vibrating away. Each of them stores energy. There is, therefore, an infinite amount of energy in every cell of space-time. Like degeneracy, infinity is a symptom that field theory is not the final theory.

It is interesting that both of these problems would be lifted by going to a discrete theory. The third disease of the classical picture is that there is no operational method of carrying out the measurements required for a continuum theory. If you try to measure the field at a point, you must put a test body of very small size, in fact of zero size, at that point. In order to know where it is, it must be classical. There are no classical bodies. It can be made approximately classical by making it very massive. So you must use very massive, very small test bodies to measure fields at a point, or nearly at a point.

When this theory was first worked out in the 1930s by Bohr and Rosenfeld, they were content with the result because they saw no reason for there not to be very massive, very small test bodies. We know more about the elementary particles today. There are no arbitrarily massive, small test bodies. They stop at a few thousand times the mass of a proton. This means there is a limit to the fineness with which we can measure fields.

The Smallest and the Briefest

DALAI LAMA: Are you implying that you reach a point where even conceptually you cannot divide any further?

DAVID FINKELSTEIN: That's another way of saying it, yes. There are actually three different things that might happen if you imagine measuring the field at smaller and smaller regions. The most optimistic one says that if you look too carefully, you will create a black hole and lose what you try to measure. That first happens at 10^{-45} seconds. If you look a little more carefully at the theory, you see a Compton limit, which is twelve orders of magnitude worse. The breakdown of ordinary space-time occurs long before the Planck length, long before the black hole.

With these last remarks, David left not only the Dalai Lama but also most of the rest of the participants behind. There are, as David says, good reasons for why the classical notions of space-time become untenable at the small scales he is talking about. These arise when we try to give empirical meaning to the concept of location: When and where does an event occur? To answer such a question requires a probe that can measure the place and time of the event to very high precision. If we wish, for example, to know the location to within a millionth of a meter, then the probe should be no larger than that and it must be focused precisely. At this point quantum mechanics enters in. When we constrain the size and focus of the probe object, then we will, by Heisenberg's uncertainty principle, disturb it and give it energy. The tighter the focus, the greater the energy inserted into the region being examined. In addition, since energy is equivalent to mass (E = mc²), the energy required by Heisenberg's principle will introduce excess gravity exactly as if mass were present. At some point this mass-energy could become so great as to create a gravitational singularity in the fabric of space-time, that is, a black hole, destroying all possibility of measurement. Long before one reaches this limit, which is called the Planck length limit, other fundamental limitations on measurement arise. The so-called Compton limit is one of them.

When an energetic photon scatters from a stationary electron, its wave-

length changes slightly. The amount it changes is some fraction of the so-called Compton wavelength of the electron, which is 2.4×10^{-12} meters. If the photon scatters from a proton instead of an electron, then the pertinent Compton wavelength would be that of the proton, which is 1.3×10^{-15} meters. The larger the mass of the scattering particle, the smaller the corresponding Compton wavelength. The limited types of elementary particles (the so-called mass spectrum) limit the largest mass one could use for measurement purposes, and therefore also limit the smallest Compton wavelength that can be associated with a probe particle. David and his students have done a careful analysis of the limits thereby imposed by Compton scattering effects on our ability to measure the space-time coordinates of an event and have shown them to be a trillion times larger than the limit imposed by the Planck length.

ALAN WALLACE: You've gone way beyond the horizon of being able to translate. A probability translation is taking place here, but in terms of any precise, discrete translation, forget it. . . . Experimentally, can you identify the shortest discrete duration of time? Not mathematically but experimentally because, of course, mathematically you can make a number as small as you like.

Quantum Space-Time

DAVID FINKELSTEIN: Let me move to that immediately. Let me stop saying nasty things about classical space-time and now attempt to say positive things about quantum space-time, with the understanding that the theory is yet in a very formative stage. First one must set the scale. If there is no continuum, there is probably a least unit of time. Early in the century, when the quantum theory was discovered, they speculated about the possibility of a quantum of time, and they named it the chronon. I see no need to change the name. The arguments I mentioned before indicate that the chronon is somewhere between ten to the minus twenty-four and ten to the minus thirty-six seconds. I'm sorry that the uncertainty is so great, but that's simply the primitive stage of the theory.

DALAI LAMA: You can experimentally ascertain the duration of a chronon? How do you know that a better experiment would not come up a smaller number?

DAVID FINKELSTEIN: It is very difficult to judge the limit of a theory from inside the theory. I am using classical space-time to estimate the limit of validity of classical space-time. Obviously, I have combined this with experimental information. The main information is the mass spectrum of the elementary particles, which limits how short a time we can resolve in our measurements of field theory. This range of magnitude is the Compton lengths of the known particles that one could use as test particles.

DALAI LAMA: Is it correct that the duration of the chronon is the shortest duration of any identifiable change?

DAVID FINKELSTEIN: Of a single identifiable change. One can always make smaller times by averaging. If you have a million processes and all of them take zero time except one, which lasts for one chronon, the average duration will be a millionth of a chronon.

DALAI LAMA: So, why wouldn't you call this point a state?

DAVID FINKELSTEIN: In fact, the next step in the conceptual development would be to call the point a state, and therefore I'm going to eliminate the point.

DALAI LAMA: Because otherwise there is still "reality" there.

DAVID FINKELSTEIN: Exactly right.

DALAI LAMA (*with a laugh*): Before we come to reality, we probably have to realize something else—lunchtime.

During the lunchtime conversation among the scientists, we decided to quickly complete David's presentation and then to open up the discussion to others. David wished to share some of his own research that attempted to understand forces and particles as expressions of hidden structures of a quantum mechanical and relativistic space-time. He invoked images of checkerboards and crystal lattices as aids, but through the afternoon session the translators had to interrupt continually to have David elucidate technical ideas. By the end, however, a beautiful conception of physics stood before us.

DAVID FINKELSTEIN: Having given up the negative part of classical physics —the idea of point objects and classical states—now we have to actually build an effective space-time out of what is left, which means operations. The procedure I will follow is the only one available for someone with my limited talent: to make models and see what they imply; to hold onto them where they work, and where they don't work, try to fix them.

The earliest discrete model of space and time within a quantum context might be that of Richard Feynman, who showed that if you look at the moves of a man in the game of checkers, and just add on quantum to the position, that is, to allow for complementarity, then you discover that nature obeys the same laws that the electron does, except for the fact that the game is two-dimensional instead of four-dimensional.

ALAN WALLACE: Please define what you mean by complementarity in this context.

Bohr's Complementarity Principle

DAVID FINKELSTEIN: Bohr introduced one of the key concepts of quantum theory to explain the new situation. Besides all of the classical, logical relations among predicates, there is a new one peculiar to quantum theory. The relationship between two perpendicular polarizers is classical: One is the negation of the other. The relationship between two parallel polarizers is also classical: One is equivalent to the other. The relationship between two oblique polarizers is not classical. [*See figure 3.1.*] This is where the ordinary laws break down. In practice, positioning filter A in front of filter B is different from reversing the order of the two filters. In the classical cases, the order doesn't matter. In the quantum cases, the order is essential. You can see it in the particles that come out. Whenever the order in which two concepts are verified matters, we say that the concepts are complementary to each other. That means each experiment requires a choice. Each experimenter can choose one framework or the other, but not both.

Bohr's original example of complementary concepts was the particle and the wave properties of light. He went on to speculate that if you looked at what they meant in practice, perhaps love and justice might be considered complementary.

There is a completely new way of taking a classical theory without complementarity and correcting that mistake. The resulting theory could still be totally wrong, but it's a way, at least, of making sure that complementarity is obeyed. Namely, there is an arithmetic of directions in which, by adding two directions that are not complementary, you make something that is. It is as if one could say that the sum of the parallel and the perpendicular directions is the oblique direction. This is what Feynman did as a matter of routine in studying the motions of a game of checkers. He allowed for every pair of moves a player could make, and also a hypothetical sum that lies between them. He found that this theory strongly resembled the theory that we ordinarily use to describe how an electron moves in what is supposedly the continuous space-time of Einstein. The role of the speed of light, which is so important in relativity, is provided by the chessboard itself. The lines along which a bishop moves in the game of chess (diagonally) are the lines that a photon follows in special relativity. The geometry of space and time is constructed out of the cells of the underlying space-time network.

The world cannot be a classical checkerboard. A checkerboard has special directions. The world has no special directions. But the world could be a quantum checkerboard because all the missing directions could just be complementary mixtures of the directions of the classical checkerboard. This idea, that the roundness of space is entirely a quantum effect, was put forward by von Weizsaecker in the late 1940s and has been propagated through the work done since then. I'm trying to pursue it still further in my own work. It's easy to make a four-dimensional chessboard, the fourth dimension being time.

The Isotropy and Curvature of Space-Time

DALAI LAMA: Are you saying that space is round, in and of itself? Or that space is round in some other sense?

DAVID FINKELSTEIN: It is round in the sense that all directions are the same. The real meaning of this roundness is that any experiment that works continues to work in exactly the same way if you turn all the apparatus.

DALAI LAMA: Why wouldn't the scientist draw the conclusion that space is infinite rather than curved? The effect would be the same.

DAVID FINKELSTEIN: I'm describing one particular model, which is, in fact, an infinite space-time. I did not mean that space-time is curved. It is, to begin with, flat, just like a chessboard. Sorry, I shouldn't have used the word *round*; it's confusing. The technical term is *isotropic*. It just means the same in all directions.

DALAI LAMA: Yes, I see.

DAVID FINKELSTEIN: We'll come to curvature in a moment. It turns out that simply by turning on superposition, in much the way that Feynman did, one can turn this very angular four-dimensional chessboard into something that has all the kinds of symmetry that special relativity requires. There is no preferred time direction. There is no preferred space direction. It follows the laws of special relativity.

But then the phenomenon of curvature needs to be expressed in this language. General relativity, which deals with the curvature of space-time, is actually the father of a whole line of modern theories. The so-called standard model of the strong, weak, and electromagnetic interactions are all modeled on the basic structure of Einstein's theory of gravity. The basic idea is that not only the laws of nature but even the concepts of nature must be local. There should be no assumption that it's possible to compare, for example, a direction at this point of space-time with a direction at another point of space-time without actually carrying it from here to there along a definite path. In the geometry of Euclid, it is supposed to be meaningful to say that two lines are parallel, even though they are very far apart. For the physicist, the question comes up, How do you know? In Einstein's way of thinking, it is impermissible to introduce as a fundamental concept such comparison of remote things because we are not in both places. You must describe the procedure by which you take a direction here and carry it there, step by step, and see if the lines are indeed parallel. This principle of locality is the second cornerstone of modern physics, along with complementarity, which is the first. The struggle is to bring them together, as it were, to marry Heisenberg and Einstein, who did not like each other.

Particles and Forces in Quantum Space-Time

David has explained the concept of complementarity through the phenomena of oblique polarizers. Locality requires that we define physical concepts in terms of local operations. These two concepts—locality and complementarity—become the foundation stones for a consistent view of quantum space-time. The first concept we have from quantum theory, the second from the theory of relativity. To bring these two theories together (which has never been done successfully) to form a discrete crystal-like model of space-time, David changes our view of space-time from a lattice of points to one composed of quantum cells, which are the fundamental units out of which space-time is constructed. But space-time is not perfect; the crystal has flaws. It turns out that these flaws can be understood to be the properties of fundamental particles and forces between them. Although David quickly asserted that this is only a model, and one that could well be utterly wrong, nonetheless he found it to be a beautiful way to capture the deepest principles on which our world is constructed.

DAVID FINKELSTEIN: Locality and curvature express themselves in a crystal through defects in the crystal. If, for example, a plane in the crystal is missing from a certain point on, a half plane of atoms is missing. Then a path that would otherwise close in a good crystal will no longer close. You go around a loop and you don't come back home again. If the crystal is missing a whole sector and is drawing in on itself to close up the hole, then when you carry a cell around a loop in that crystal, it will come back turned. In such discrete models of space-time, we rely on defects in the vacuum crystal to account for all the phenomena like curvature and ultimately all of the forces. The minute you make such a model, it is important to ask what happens to a piece of it when you carry it around a loop and bring it home again. My first shock in this research was to discover that the things that happen to a cube when you bring it around are practically the things that can happen to a quark in physics when you bring it around a closed loop. There seems to be an intimate relation between what are called the internal degrees of freedom of the quark and the space-time structure.

This model may be totally wrong, but within this model I can show you which kind of defect results in light, which kind of defect results in the strong gluons or any of the known forces, like gravity. All the forces turn out to be modeled rather well within this discrete model. That does not mean it's correct, but it encourages me to carry it further.

These results are very robust. They do not depend on the detailed structure of the crystal, only on its resemblance to a chessboard. It leaves open the question of what goes on in each square, so to speak. The question then arises, What is actually happening? What are the actions? Again, all I can do is make a simple model. The oldest model of a network is that of the natural numbers created by the Italian mathematician Giuseppe Peano in the last century. He started with nothing. He took that to be the model for the zero point. Then he took the set whose only element is nothing and took that as the model for the number 1. The set whose only element is the number 1 is then the model for the number 2. This process of enveloping in deeper and deeper sets Peano used to model the passage of time. This envelopment can also be regarded as a transition from a level to its metalevel. By the metalevel I mean the level which knows the level. For example, in mathematics, when we discuss a language, the language in which we talk about the language is called the metalanguage.

The remarkable thing about Peano's model is that the points of his time are generated by this one relation between them. It is a model in which the points do not preexist but are generated each from the previous one in a dynamic way. It's not difficult to generalize this to four dimensions. In this model, the whole world is a pattern of transitions from system to metasystem, or from knower to known. The metasystem is that which knows the system: the experimenter, the apparatus, everything around the system that is involved in forming the concept of the system. Physics until now has been very flat. There is the system that we talk about, and we repress the metasystem completely, as has been emphasized earlier. This can't go on. In quantum theory, for the first time, we see it's impossible to avoid reference to the metasystem.

David's final remarks returned us to the previous discussion of the preexistence of properties for quantum objects. Under Anton's guidance, we had already encountered problems with the idea of a classical world whose objects possessed objective properties. Here David was invoking a similar claim for space-time itself. Points in space-time do not exist; rather they and the whole world with them are to be understood as a pattern of transitions or actions. These actions move us from system to metasystem, from knower to known. David sees the knower as being at the level of the metasystem, and he was passionate about the inclusion of this lost feature of the world, the knower. Classical physics had left the knower out of its account and so never rose to the metalevel. Its account was entirely at the level of supposed objective reality. Quantum mechanics and relativity have called all this into question. The task before us is to frame a consistent account that embraces the far more subtle and complex phenomena and principles of the new physics as depicted by David and Anton.

Absolutes within a Relativistic World

In the following exchange, the Dalai Lama probed insistently for absolutes. Within Buddhist philosophy, one is constantly alert to the habit of the mind that will reify a useful construct, converting it thereby into an absolute when, in fact, it should properly be conceived as an important but conventional way of understanding. Many of our prejudices, such as racism, are grounded in such mistaken reification. We take a collective opinion as fact. As difficult as it is to free oneself of these false absolutes, it is far more difficult to examine critically some of our most basic conceptions, for example, our conception of space and time. These seem absolute, and yet, what is their fundamental status?

DALAI LAMA: Within the context of physics, certainly, there are theories about space. When you conceive of space, whether it's in terms of relativity theory, Newtonian mechanics, quantum mechanics, or any other theory, do you define or conceive of space purely as an absence of something? In Buddhism, when we speak of noncomposite space, this is conceived of as a sheer absence of any obstructing entity. And you stop there. You don't say it's an absence of that and then go on to make affirmative statements about the nature of noncomposite space. When you speak of composite space, which is the second type of space within Buddhism, then more may be said. But when you, as physicists, think of space, is it not true that you go beyond merely negative statements of the absence of something? Don't you also make affirmative statements, for example, that it can be curved?

ANTON ZEILINGER: The way we view it is that space is only defined by the bodies in space. The bodies define the space. The notion of space in itself without stars and so on is an empty concept. It does not make sense for a physicist.

DALAI LAMA: Please define it. Assume you have a body of space with material objects in it. Now how do you define space?

ANTON ZEILINGER: Well, once I have the objects, I can start talking about distances. I can start talking about the path I take in the space because I can refer to the bodies. If I take all bodies out of space, then there is nothing left. If I can't refer to the bodies, there is no motion, for example.

DALAI LAMA: Perhaps there is a parallel with time as well—that you don't see time in the abstract but always time between two events. It has a kind of imputative status in reference to something that is not time. You have one event and another event, and on the basis of that you impute the notion of time. Just as you have two physical entities, and on the basis of that you impute space, but not without them.

ANTON ZEILINGER: Absolutely, because space is a way to say where something is. Time is a way to say when something happens. That can only be said in relation to something else. In itself it has no meaning.

DALAI LAMA: In one of the Abhidharma texts, which is the Buddhist version of classical physics, there is a concept of composite space. This is defined in terms of that which captures the light during the day, when the sun is shining, and becomes dark during the night. It's a medium. Do you regard time as absolute or relative? I have heard Westerners articulate both views.

ANTON ZEILINGER: I would say time belongs to those things of which we have learned that we can only make relational statements. So, I would say there is no absolute time. It's only in relation to something else.

DALAI LAMA: Then is it safe to say that, from the scientific point of view, no notion of the absolute is tenable?

ANTON ZEILINGER: Most likely.

DALAI LAMA: This would be in accordance with the thought of Nagarjuna, who does away with absolutes altogether.

Anton's refusal to respond in definitive (i.e., in absolute) terms drew a laugh from everyone. As David and I joined the discussion, I pointed out the thermodynamic basis for "time's arrow." It is interesting that David at least partially defended the use of absolutes in physics, thus standing in disagreement with both Anton and Nagarjuna.

ARTHUR ZAJONC: I agree with my colleague Anton, on the one hand. On the other hand, when you hear people say that there is a direction to the flow of time, that they have a sense of earlier and later times, what they are usually referring to is based on thermodynamics. They are referring to the way in which matter evolves from a state of order to a state of disorder, a so-called increase in entropy. If there were no matter in the universe and one were in empty space, then of course it would not be possible to use this as a guide. In an empty universe, the sense of space or time becomes meaningless, by and large. Would you agree with that, David? You're the most expert in this subject area.

DAVID FINKELSTEIN: You've been speaking from so many viewpoints. Everything you said is true at a certain point of development, but development hasn't stopped there. For example, it was true in the beginning of the century that concepts of space and time and space-time were highly relative, but one of Einstein's contributions was to make space-time practically a material system in its own right, to give it independent existence. Even without bodies you can have gravitational waves, which are waves of curvature of space-time, propagating through space-time. This conflicts with much of classical thought, but it is the best we can do today. Nonetheless, I agree with Anton, that very likely the idea of an absolute space-time will go the way of many other absolutes. But at present physics is full of such absolutes, and one of them is space-time, as opposed to space or time.

DALAI LAMA: Perhaps you can define what you mean by absolute? When you say there are no absolutes or that space-time is now regarded as absolute but may turn out not to be absolute, what exactly do you mean by absolute?

DAVID FINKELSTEIN: In any physical theory, you begin with the description of the system or of some properties of the system. But the theory never stops there. You begin with a description from one point of view, and then you explain how this is related to other points of view, for example, "This is how the relativity theory is associated with that physical theory." So each theory carries with it a concept of an absolute or physical objectivity. If, from one observer's description of an entity, you can compute the description of all other observers, then we call it objective. When everyone can translate their descriptions into each other's language, we say we have an objective entity. In physics, objectivity has become intersubjectivity. Points of space no longer have objective existence. Time no longer has objective, absolute meaning. An interval that one person says is space, another person, we realize today, will say has the element of time in it. An interval that one person says is purely an interval in time, another observer will say has a spacelike component. For example, when I took the train from Delhi to Patankot, I stayed in one cot for the whole trip. So the question could arise as to whether I moved in space. From the point of view of the train conductor, I was a good passenger and I did not move. Of course, from the station manager's point of view, I was a good passenger because I did move. Otherwise I would ask for my money back. In one case there is a spatial component, and in the other case there is not. But everybody agrees that there was a certain event, which was my departure from Delhi at a certain place in space and time, and another event, which was my arrival in another place in space and time. Those, at present, are still absolutes. From one's person's description, another person's description can be computed. The number of absolutes is steadily decreasing. The group of relative entities is getting larger and larger.

DALAI LAMA: So you are defining absolute in terms of objectivity.

DAVID FINKELSTEIN: Right.

The Observer in Einstein's Theory of Relativity

The shifting nature of space and time becomes especially dramatic in Einstein's special relativity. In this theory, now well supported experimentally, lengths are foreshortened in the direction of motion, and time runs slow for a moving clock. These features of modern physics are difficult for us to fathom. Our introduction of these ideas to the Dalai Lama was received with some skepticism and many questions. Buddhist philosophy often concerns itself with sources of delusion, including the case of illusory motion. For example, in the case of a boat drifting on a river, how does one determine the true motion of the boat? Einstein began with this same problem but then showed that for the laws of physics to hold, our ideas of the nature of space and time must change. It fell to me to make the first attempt to explain the argument of Einstein.

ARTHUR ZAJONC: Your Holiness, you were interested in the role of the observer in modern physics. We talked before about the role of the ob-

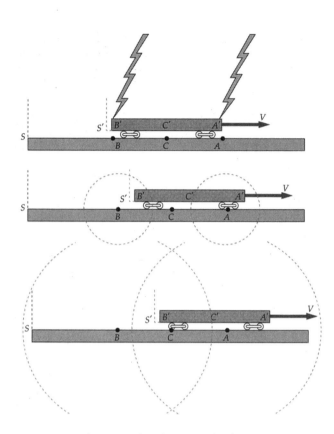

Figure 3.2 Lightning strikes the two ends of a moving train car. Judged from the embankment, they strike at the same instant. However, when judged by an observer on the train at position C', they do not strike simultaneously. Since the train is moving to the right, the lightning flash at A' reaches C' first (middle picture) and only later does the wavefront from the rear of the train reach C' (last picture). Since neither frame is privileged, this shows the relativity of simultaneity. Source: Paul A. Tipler, *Physics for Scientists and Engineers.* © 1999 by W. H. Freeman and Company. Used with permission.

server in quantum mechanics. It's very important also in relativity. I would like to show you something about the observer's frame of reference and its importance to relativity. Consider the following situation. There is a train moving along the tracks. There are two observers. One observer is in the train, in the very middle of the train car. There is another observer sitting quietly at the side of the railroad track, stationary relative to the earth. These two observers are indicated by C' and C in the diagram. [*See figure 3.2. In what follows the primes always indicate the frame of the train.*] Now suppose it happens by a miracle that two lightning flashes come down and strike the two ends of the train. In the reference frame of the observer C, who is sitting quietly on the side of the tracks, he judges these two lightning bolts to hit the ends of the train at exactly the same instant. It could even be that some significant event happens. Maybe there are children to be born, or tea is to be served when these two lightning strokes hit the ends of the train simultaneously. Now, this disturbance, the event of the lightning strike, creates a wave of light, a signal of light that emanates from the two positions A' and B' at the ends of the train. This is indicated by the two circles expanding away from the ends of the train. But now notice, Your Holiness, that the circle on the right will reach the observer C' before the circle on the left will reach that same location.

The train is in motion to the right, as indicated by the arrow. So you can see the train moving, moving, moving; when it reaches this point, the observer C' will see the lightning flash. When sometime later it reaches this point, the observer C' will see the other lightning flash. So you have two different descriptions of the same situation. For the observer sitting quietly on the side, he sees the two lightning flashes arrive at the same moment. For the observer at the center of the train, he will experience a flash first from the front of the train and then later from the back of the train. In other words, normally we think of time in such a way that we can say quite objectively that events either happen at the same time or one happens and then the other happens. But here we can see that the sequence of events depends on the state of motion of the observer. Events can happen at the same time, in one frame of reference. The same two events can happen first in the front, then in the back, or first in the back, then in the front. This is called the relativity of simultaneity. This is one of the first discoveries of Einstein's theory. Once again, we see that certain things about our world, which we take to be fixed and firm, or absolute, begin to move and become more relative. This is an example of one of the factors that undermines ideas of the absolute and which emphasizes the role of the observer.

DALAI LAMA: Can't one defend the classical notion of time and simultaneity here by saying that the perception of the observer beside the track is the correct one because the perception of the person in the carriage is deluded by the motion of the train? In Buddhist epistemological discourse, we talk about different types of optical illusions and where the sources of illusion lie. In some cases, it could be a defect in the organ of the eye. In some cases, it could relate to the environment. For example, when you are in a moving boat, you see all the trees on the bank moving. Another example given of a source of delusion is that someone who is very angry sees things red. In this case, the immediately preceding mind-state actually distorts perception. Similarly, this morning we talked about how an individual's expectations, purpose, and concepts affect the perception. From the Buddhist point of view, these could be seen as a source of illusion, lying in the immediately preceding consciousness.

ARTHUR ZAJONC: It is very natural to immediately question whether there is some way to decide by deeper analysis which point of view is privileged or which one is in error. The answer in this instance is no. There is no deeper analysis that either observer can make of his own situation that would show that they are in motion.

DALAI LAMA: So a crucial thing to be deduced is that motion is relative—that no motion is absolute?

PIET HUT: If this story were to be told with sound instead of light, with somebody blowing a whistle instead of lightning striking, then it would be an illusion. In this case you could not conclude that the timing is dependent on the observer because sound travels through the air. The air stands still for one person and moves for the other person. But light is special. Light has no medium. For both observers, light is equally fundamental. It is not a matter of illusion. That is the main difference concerning the relativity of motion.

ANTON ZEILINGER: Maybe I can help clarify. In the story that Arthur presented here, it seems that the station is a special system. It's at rest and the train is moving. You could tell the same story with two spaceships out in space. Then it would be impossible to say that one is at rest and the other is moving. And you would arrive at exactly the same conclusions.

These remarks sparked a discussion in Tibetan between the translators and the Dalai Lama. The difference between a delusion, which can yield to more careful analysis, and the genuinely ambiguous situation one has in relativity is subtle but essential for modern physics. The translators asked for clarification.

ALAN WALLACE: You have two spaceships in relative motion. You can posit that one is stationary in relationship to the other one's moving. You can switch it around and say the other is stationary and this is moving. Or you can assume that they are both moving. But there is no way to say what's really going on independent of frame of reference.

THUBTEN JINPA: You can turn off the battery of one.

ARTHUR ZAJONC: The battery is not necessary for motion. They are just coasting. We should do the next Mind and Life Conference in two spaceships. We'll talk with NASA about it.

DAVID FINKELSTEIN: We are illustrating two principles at once that Einstein carefully made explicit. One is that there is no operational conception of motion. One can only speak of relative motion. This was already emphasized by the Italian scientist Galileo in the seventeenth century. The new discovery, which is equally important, is that, nevertheless, all observers, no matter how they were moving, will agree on the speed of a particular light wave. That's an astonishing paradox from the point of view of classical physics. One could imagine that one could catch up with a light wave, and then it would seem to be at rest. The whole point is that one cannot catch up with light waves. No matter how we move, they always seem to travel at the speed of light relative to us. This requires a revision in our notions both of space and time. That was the point of the story, to demonstrate a revision of the conception of time. It does not mean you have to give up the idea of a medium and imagine that waves are traveling in nothing. As my teacher Weisskopf was very happy to point out, you should say rather that the vacuum is a very special kind of medium, which is such that you cannot detect the rate at which you move through it. Speed relative to this medium is not operationally defined.

ANTON ZEILINGER: I would like to conclude with one thought. The famous Danish physicist Niels Bohr said there are two types of truths: simple truths and deep truths. He said that a simple truth is a truth where the opposite is not true. A deep truth is a truth where the opposite is also true.

DALAI LAMA: There's something to that.

The search for deep truth, for deep insights into the nature of the world around us, has gone on for millennia and in all cultures. In Asia, as in Europe and the United States, we have used our best reasoning and our most careful observations to help us discern the enduring structures of reality. Through quantum physics and relativity, but also through careful philosophical analysis, that which one age took to be insight was later shown to be only partially correct. The lesson from this history of science is not that all efforts at understanding the world are futile but rather that our understanding must be dynamic and contextual. In Buddhism, as in physics, many factors suggest views such as those advanced by David Finkelstein and Anton Zeilinger. Although they agreed in many things, they also differed in important particulars. David dethroned objects and their atten-

dant states but sought a view of relativistic and quantum space-time that could act as the deep operational structure of reality. I sensed that Anton was more reticent, adopting a cautious Copenhagen stance and always returning to what could be measured or investigated experimentally. The theoretical pictures of David, although alluring and even brilliant, always needed to be grounded on the facts of observation. These facts themselves constitute our reality more than anything else.

On Wednesday, I wanted to take up these latter issues with the Dalai Lama. What should be our attitude to theoretical entities remote from view? How should we handle the qualitative observations of the senses versus the quantitative measurements common to modern physics?

4

Buddhist Views on Space and Time

*Following each morning's presentation, the participating scientists gath-
ered over lunch and suggested topics for the afternoon discussion. The
hope was for a more spontaneous and wide-ranging conversation, one that
would include responses from the Buddhist perspective to those issues
brought up in the morning. On the afternoon of the second day, after the
completion of David Finkelstein's presentation, I asked the Dalai Lama to
tell us something about the Buddhist understanding of space and time. His
presentation sketched for us the subtle distinctions that Buddhist philosophy makes between our subjective experience of time and time's own nature. It also led to an animated discussion about the origins of time, the big
bang theory of the cosmos, the Buddhist notion of space particles, and the
manifold sentient worlds that Buddhists think span the universe. Many of
the themes treated in this chapter foreshadow our later discussions of modern cosmology, which occurred on day four and which we will return to
then. The Dalai Lama began with remarks on the nature of time.*

DALAI LAMA: For many, but not all, philosophical schools within Buddhism, a distinction is made between what may be called substantial entities and imputed entities. A substantial entity is something that can be
identified in its own right. You could say, for instance, "Here's a bottle."
You don't have to look at something else apart from the bottle in order
to identify it; you just look at the bottle and there it is. You can put your
finger on it, metaphorically speaking. That may be also true of mental

phenomena. There is another class of phenomena called imputed, or designated, entities. You can't simply put your finger on an imputed entity, in and of itself. It can only be designated in relationship to something other than itself. Time is a perfect and archetypal instance of an imputed entity. You can't simply put your finger on it and say, "There's a chunk of time." You designate time on the basis of some process of change, something else that itself is not time.

GEORGE GREENSTEIN: Time is known by the fact that things change?

DALAI LAMA: No, not exactly. Time is a designated entity because it is identified on the basis of something else that itself is not time. When you look at the clock, you say time is passing. What you are looking at is the second hand. The second hand is not time; but based on seeing the second hand move, you say that time is passing, that five seconds have just passed. That is a valid designation, but you are not putting your finger on the time itself. You are putting your finger on something that is the basis for the designation of time.

In the context of this school of thought, time is always designated upon a continuous sequence of events, which means that it is designated solely in relationship to changing composite factors. Time is never designated upon something that is noncomposite such as noncomposite space.[1] But if, in this context, you ask what the nature of time is, it does exist. You can't say that it has no nature whatsoever. It's not merely a conceptual fabrication. It does exist, but consider what happens if you try to identify its nature.

For a substantial entity such as the bottle, you can describe the stuff that it's made out of—plastic. If we ask a similar question about time—what's the stuff that time is made out of?—you can't answer because time itself is designated upon something else that is not time. Consider then whether time is a permanent or an impermanent phenomenon. Time itself is not permanent; it's impermanent; but what kind of impermanent phenomenon is it? Is time itself a physical phenomenon? No. Is time a mental phenomenon? (We know a lot of mental phenomena are impermanent.) No, time itself is not a mental phenomenon. So, within the basic Buddhist phenomenology of the structure and classifications of reality, time fits into a third classification: impermanent phenomena, neither physical (in modern terminology, they are not of the nature of mass-energy) nor mental, of the nature of cognition. Technically, this category is called impermanent, nonassociated, composite phenomena. Nonassociated means it is not included in the physical or in the mental. Nevertheless, it is impermanent.

Noticing that some of us were looking a bit confused, the Dalai Lama laughed and remarked, "I, myself, am confused." Then he plunged ahead into a discussion of space.

Time and the Buddhist Classification of Phenomena

DALAI LAMA: The classifications of phenomena also include affirmative and negative phenomena. The class known as negative phenomena can be established only by the process of negation. The example I gave earlier of noncomposite space, which is established on the basis of the sheer absence of obstructing contact, is an example of a negative phenomenon. It does exist, but it is defined as the absence of something, and that absence does exist. But there is another class of affirmative phenomena. These can be established without the process of negation, purely through affirmation. Time is an instance of such an affirmative phenomenon.

Noncomposite space is established negatively and is therefore, in itself, not changing or impermanent. There is another class of space that is accepted by the Vaibhasika school, which was referred to briefly yesterday. This "composite" space is something that is conditioned—it changes as a result of different causes and conditions, so it is an impermanent phenomenon. Moreover, it is a physical phenomenon and it is a medium. It is, in effect, the medium of light and the medium of darkness. It is also a focal object: It can be perceived visually and is seen as part of the condition for visual perception of an object. Another word for it is *intervening appearance*. For instance, as I look out at George, I can see the space, the intervening appearance, between us. This is sometimes called luminosity, but it really refers to space and not light. This is something that is conditioned.

This type of space is accepted by general consensus in Buddhist philosophy, from the most basic Vaibhasika School[2] to the most subtle Madhyamika. It is also accepted in the Kalachakra,[3] which is the highest expression within Tibetan Buddhism of an ontological system and which is also a phenomenological system. But in the Kalachakra, the terminology used is "space particle," implying that space consists of particles. This opens an interesting dialogue because it clearly implies a quantization of space, not a mere negation and not something unconditioned.

The Dalai Lama then moved beyond traditional Buddhism to a more general neo-Buddhist perspective in order to connect his own understanding of Buddhism with contemporary astrophysical models of the early universe. In particular, he showed an interest in the big bang model of cosmogenesis. The "space particles" of the Kalachakra philosophy suggested to him a way of thinking about the very long-term history of the cosmos, including times before the big bang. This point is important in Buddhism since it does not accept the idea of creation out of nothing, which is common in the Judeo-Christian West. If one accepts the big bang, the Dalai Lama suggested, one believes that it must have been a transition event from an even earlier state of the universe, perhaps involving a series of such arisings and passings of the material cosmos. Some Western cosmologists have suggested such a model of an oscillating universe. The Dalai Lama went on to contrast this discussion, which entails vast eons of time, with the Buddhist analysis of the smallest unit of time as subjectively determined through human experience. The phenomenological emphasis of the most advanced Buddhist philosophical systems is another aspect that we will return to repeatedly.

DALAI LAMA: In a different conceptual framework, a neo-Buddhist perspective, space particles would be something that existed prior to the big bang. It would not be possible for such a transition to take place if there were no space particles. They are the basis, the source, of all phenomena that arise in the universe. Traditional Buddhist cosmogony describes how the universe evolves out of times of sheer emptiness of space and then eventually dissolves back into space. It is an oscillating universe. That's why we can say, "before the big bang"—even though that is a modern Western concept, there is an analogous concept in Buddhist cosmogony.

Concerning classifications of time, there is a general consensus that there are three times: past, present, and future. What is the nature of each of those and the relationship among them? How do they differ? Different philosophical schools have different interpretations. It's rather complex—I don't remember all of them, and I can't very well ask you to wait here while I go read up on the texts. But within the Buddhist concept of time, there is an appreciation of the divisibility of time. In the context of cosmology, there is a discussion of the duration of the formation of the universe, eons and great eons. On a smaller scale, there is a distinction between two key divisions. One is the shortest possible time of a complete act. Examples given are the moment that it takes to blink your eyes or snap your fingers or say, "Ah." These are considered to be the shortest moment during which an ordinary person can perform a complete act, so it is anthropocentric. There is also a concept of the shortest divisible unit of time. One school of thought in Buddhism maintains that it is one-sixtieth of the duration of a finger snap. There are others who argue that it is one-365th of the duration of a finger snap. Of course, these are very crude divisions by comparison to modern mathematical divisions of time.

Of course, one can pursue further the analysis of the nature and divisibility of time, but eventually you reach a point where the notion of time gets lost. This applies not only to the nature of time but also is true for the analysis of all phenomena. In the Madhyamika school, such investigation is seen as stemming from a dissatisfaction with appearances, with the level of pure observed phenomena. All of this takes place within the context of what is known in Buddhism as conventional discourse, where the validity of conventions is seen as a measure of judgment.

Madhyamika Philosophy and Conventional Reality

Occasionally, Alan Wallace would step out of his role as translator to comment on some aspect of the topic under consideration. Here he comments on the use of the difficult phrase conventional existence.

ALAN WALLACE: A crucial point here is that, in Buddhism, when we say *conventional*, we do not mean whimsy. In English, *convention* has the whimsical notion of something that people say arbitrarily. In Buddhism, something that is described as *conventionally existent* is taken very seriously and contrasts with something that is thought to exist absolutely.

DALAI LAMA: If you assume that time—for example, this shortest unit of empirical time—exists absolutely, and then you really subject it to analysis, you find nothing there. Therefore, even though a shortest unit of time is posited, it is posited conventionally, not absolutely. As this is true of time, so it is true for everything else. If you subject anything—space, time, matter, whatever you like, even the mind itself—to a certain type of close scrutiny, looking for its actual nature independent of other phenomena, it will dissolve under analysis every time you look for it. Then, if you agree that it is empty of inherent existence, and you try to seek out the very nature of that emptiness, the emptiness itself is not to be found. Therefore, emptiness itself has only a conventional and not an absolute existence.

ALAN WALLACE: To translate *emptiness* as *the void* is a terrible translation, very misleading. The term *void* implies something that is really there to be found.

The Five Elements

DALAI LAMA: The Buddhist classification of the material world includes five elements: earth, which refers to solidity; water, referring to fluidity; fire, referring to heat; air, referring to motility; and space. (Space in this context is a conditioned phenomenon, not the permanent, noncomposite space.) Then, still within the physical world, there is a myriad of phenomena that are said to be derivative of these elements. The classic list of phenomena derivative of the elements is visual form, sounds, smells, tastes, and tactile objects. David, using this kind of matrix of classification, would light itself be regarded as an element, a fundamental constituent, or is it a derivative of the elements, as sound is?

At this point the Dalai Lama invited David Finkelstein, who was still sitting across from him, to think about light within the classification scheme offered by Buddhist philosophy. David answered cautiously in terms of his own view of physics, which avoids permanent entities in favor of operations or acts.

DAVID FINKELSTEIN: Are the elements in some sense permanent, or can they be momentary?

DALAI LAMA: They are all momentary. None of them are permanent. A crucial characteristic is that they are all constantly in a state of flux.

DAVID FINKELSTEIN: One could speak of elemental acts. . . . I will answer now from the framework of my very speculative theory, which may not last the week. *[David laughs.]* Neither the photon nor any of the other quanta are elemental. They are patterns in the structure, which also manifests itself as space-time structure.

DALAI LAMA: This afternoon you spoke of the quantum continuum or the quantum space-time. Would it be correct—again, inviting you into the Buddhist framework—to say that, in fact, there is only one element, the quantum space-time? Is everything else, all forces, space, time, light, matter, simply derivatives of the quantum space-time?

DAVID FINKELSTEIN: I was using the word *element* as that of which something is composed, in the way that water is composed of molecules. If that is our understanding, then, yes, everything else is indeed a patterning of the structural form of the quantum space-time element. This might be seen as an alternative to Einstein's program of a unified field theory, for example.

DALAI LAMA: Just as you see quantum space-time as the source from which all other things are derived, so in Buddhism there is a definite sequence in the cosmogony for the emergence of each of the elements. From space (and this is conditioned space), air emerges as motility; from air, fire emerges as thermal energy or heat; from fire emerges fluidity, liquids of all sorts, represented by the element of water; from water emerges earth, or solids of all sorts. In this way, you could say that the latter four elements are all derivative of space, which is the fundamental one. This is accepted by all schools of Buddhism and not just limited to one esoteric doctrine. *[Turning to Weiming]*: What is the Chinese theory concerning the five elements?

TU WEIMING: It would be better to render them as phases rather than elements. They are dynamic, transformative moments rather than simply discrete elements. They also emerge out of a notion called the Tao, which is ineffable, cannot be fully comprehended, and yet it seems to have an inexhaustible supply of potential and authentic possibilities.

DALAI LAMA: That is closely related to the *Dharma*, the spiritual path, is it not?

TU WEIMING: Absolutely.

DALAI LAMA: Whereas the Chinese theory is deeply related to spiritual practice and a spiritual view of the universe, the Buddhist theory of space and the derivative elements is purely a physical description. It's Buddhist physics and not Buddhist soteriology, or spiritual practice.

TU WEIMING: The understanding comes from the meditative, and intersubjectively confirmed, understanding of ultimate reality. That view of physics is closely linked to many, many forms of spiritual exercise.

The Meditative Experience of Space and Time

DALAI LAMA: There are two different perspectives in the Buddhist discussion of time and space. The one I just described is presented in the Buddhist texts as a purely objective theory about the nature of the physical universe—objective in the sense that it need not be experienced in a meditative state but is simply what's there.

There are also modes of experience or phenomena that emerge through the power of a contemplative's own transformed mind, and they don't exist without that. If you empower your mind by various contemplative practices, a certain realm of reality arises through the maturation of your contemplative insight. Take the example discussed in some Buddhist texts of how meditators in highly evolved states are able to experience eons shrunk into a single instant of time, and also are able to stretch a single instant of time into an eon. From a third person's point of view, what the meditator experiences as an eon is seen only as a single instant. The phenomenon is subjective, unique to the meditator alone. In fact, we need not invoke the example of highly evolved meditators. From our own personal experience, we find that if we have had a very good sleep, say for five hours at a stretch, we can wake up feeling as if we had slept for only five minutes.

In Buddhism in general, the discourse is often divided into three parts. One is the presentation of reality as it is. The second is the mode of procedure on the spiritual path, based on an appreciation of the nature of reality. The third is the description of the fruition of the path. This format seems to suggest that it is important for Buddhists to clearly distinguish between reality that can be intersubjectively verified or disapproved and a meditator's own perspective, which may be unique to an individual and not intersubjective. It is important not to mix the different types of discourse. The famous Tibetan thinker Tsong-khapa said that unless we are clear on this distinction, we will have a very muddled understanding of the nature of reality. All the discussions about the nature of reality that take place in the scientific context should be incorporated within the first type of discourse.

Concerning the problems surrounding the use of the word *reality*, in Buddhist discourse one can talk about reality or about states. It's not that we have problems with the notion of state itself. Rather the problem is the notion of state as an intrinsic reality. So long as we are aware of that danger, and aware that we are using language in a very relative sense, then we have no difficulty with using such words. We affirm the nominal status of states but reject the reification of them.

In Buddhism, there can be two or three types of discourse concerning the nature of reality, and it is very important not to confuse them. For example, the normal discourse of science concerning space, time, energy, electrons, photons, and the like forms one type. We reason on the basis of experience about the way the world is and posit entities such as photons and electrons. In Buddhism, the comparable discourse concerns time and the five elements. Yet Buddhism denies the objects of that discourse any intrinsic reality, be they electrons or the element water. Buddhism treats them as nominal in character, or conventional, *to use the term Alan defined for us. This is not to deny the utility of the concepts, nor even their "relative" reality, but to regard them as having no intrinsic existence.*

The Dalai Lama contrasted this type of discourse with another that is more closely connected with experience itself. We experience existence through our senses, but within the contemplative traditions of Asia it is common to undertake spiritual practices designed to transform the mind and experience. A second form of discourse is based on the personal experiences of the experiencer or meditator. Since experiences are dependent on the maturity and capacities of the practitioner, they may well be particular to him or her; the Dalai Lama used the phrase "unique to the meditator alone." Speaking from the perspective of Chinese philosophy, Weiming suggested that meditative understandings can be intersubjectively confirmed, which moves them beyond purely personal experience.

In Western science and philosophy, I often sense that, indeed, different types of discourse are mingled without much attention. We physicists speak quite easily about all manner of theoretical entities such as photons and the like. It is eminently practical to do so. However, we usually don't wish our descriptions of them to be taken literally; we know that extreme caution is required if we are pressed to talk about "reality." Some philosophers of science (the antirealists, for example) have taken this point very far, denying reality to the theoretical entities of physics.[4] Among scientists and philosophers of science, there are many schools of thought concerning the "reality" of theoretical entities. We will return to this point on day three. The second style of discourse in science is more empirical in character and stays much closer to observations. This position also comes in many variations, but Anton Zeilinger is a good example of those physicists who often stay close to the observables. Buddhism apparently recognizes the value in both modes of discourse—theoretical and observational—and grants each its own significance without unduly reifying theory or minimizing subjective experience.

The Origins of Space and Time

After our afternoon tea break, we reconvened and took up a discussion of the origins of space and time in both Buddhism and Western astrophysics. Anton began the conversation with an allusion to certain cosmological models in which space-time arises in the course of the creation of the early universe.

ANTON ZEILINGER: Some physicists, including myself, believe that space and the universe emerged together—that the universe started out very small and defines its own space. When the universe gets bigger, there is more space. If the universe were to collapse again, space would disappear again. From that viewpoint, we don't need the concept of space existing as a precondition.

DALAI LAMA: Within the Buddhist framework you have an oscillating universe, compatible with the idea of multiple big bangs. But you need some stuff out of which a big bang would take place, hence the space particles. Out of the space particles comes the motility, the kinetic energy, the air element. From that comes the thermal energy, and perhaps that would be where a big bang takes place. But if there isn't even any space prior to the big bang, if the formation of space and the big bang are simultaneous and space is increasing as the universe is increasing, then the natural question is what catalyzed the big bang, what made it happen?

GEORGE GREENSTEIN: Is that a question about the scientific view or a question within the Buddhist framework?

DALAI LAMA: The question is answered within the Buddhist framework: Space particles are the ground from which the universe arises, catalyzed by karma.

At this point Alan Wallace broke in to explain that the Dalai Lama has said in other contexts that the space particles are catalyzed by the karma of sentient beings, which brings them into the realm of cognitive events. His Holiness then turned to Anton and questioned him.

DALAI LAMA: If you don't even have space before the big bang, then what catalyzes the big bang and out of what does it come?

ANTON ZEILINGER: My viewpoint is that the smaller the universe, the less is happening, in a sense. Therefore, the less space you need for this to happen. Ultimately you have to ask, Where is the beginning singularity? In physics, anything we say about what goes on before that is just speculation. We should not take it seriously. The idea that there are multiple big bangs is just speculation, with no evidence at all. I should not be biased about whether there is one big bang or multiple big bangs because there is no evidence either way.

Anton is clearly being cautious, not willing to speculate, whereas the Dalai Lama is continuing to apply the logic of reasoning even to the extreme conditions surrounding the big bang. Alan enters into the discussion once again, asking further questions about the oscillating universe hypothesis.

ALAN WALLACE: But if it turned out to be the case that there is sufficient matter in the universe for the universe to eventually contract, then you'd have pretty good empirical grounds for saying that there is an oscillation.

DAVID FINKELSTEIN: No, you would have good empirical grounds for saying it contracts, but when it reaches the crunch, there is no understanding of what happens next.

ALAN WALLACE: But it would be quite feasible to assume that you have a big crunch, and then quite possibly another. . . .

GEORGE GREENSTEIN: Feasibility is different from knowledge. You have no grounds for drawing any conclusion. You have no grounds for rejecting it and no grounds for accepting.

ALAN WALLACE: Of course, I'm a novice, but I've always heard that if there is sufficient mass in the universe for the collapse to take place, this would lean toward an oscillating universe rather than not. Would you not even say that much?

GEORGE GREENSTEIN: There's another possibility, which is a universe that has existed for an infinitely long time, was contracting and reached a big crunch, and then begins expansion, which goes on forever. That big crunch was the big bang.

ALAN WALLACE: His Holiness has a very straightforward question. . . .

ANTON ZEILINGER: The straightforward questions are always the dangerous ones!

The Universe: Infinite or Finite?

DALAI LAMA: You develop more and more powerful telescopes, so you can see however many billions of light years. You are seeing galaxies out there that are 15 billion light years away, isn't that right? You see more and more and more. But empirically, do you see not only that you can't see any further, but also do you see that there are no more galaxies?

With this question we confront the issue of whether the universe is finite or infinite in extent. Buddhism has a clear preference for a limitless universe composed of many "world systems" that continuously come into existence and pass away. This question will become a central one during day four.

DALAI LAMA: If you were able to see that there are no more galaxies after a certain point, that would imply a finiteness to the universe, however big. If that were the case, the Buddhists would have a problem. Buddhism asserts a literally limitless universe. When Buddhists speak of an oscillating cosmogony, of something comparable to a big bang, a development, a big collapse, a return into empty space, then the whole cycle repeating again, this does not refer to the universe as a whole. It does not refer to everything but rather to a world system. Perhaps a comparable notion would be a galaxy or perhaps a galaxy cluster, but only one certain area of the universe. So even as one world system is dissolving, somewhere on the other side of the universe another world system is emerging at the same time. It continues infinitely, with no synchronicity among them.

GEORGE GREENSTEIN: Continuous creation. That's star formation. That does happen. We can see it happening: Stars form and eventually explode or collapse. New stars form and explode, not synchronized with each other, just as you described.

DALAI LAMA: I mentioned galaxies rather than a star or solar system because the term used in Buddhism means a thousand, thousand, thousandfold world, or a billionfold world system. A world system is one with a sun, so a reasonable interpretation would be a billion solar systems, something comparable to a galaxy. In a billionfold world system, the billion systems within it arise together. Generally speaking, they arise together, develop together, and dissolve together, though not with exact synchronicity. In the meantime, there are an infinite number of other billionfold world systems, and they are evolving not synchronously. If they happen to be in phase, it's purely coincidental. In the esoteric Buddhism of Vajrayana, they speak not only of the billionfold world systems but of clusters of them — a billion billionfold worlds, and then billions of those. So in Buddhism, you have not only galaxies but also galaxy clusters and mega–galaxy clusters.

GEORGE GREENSTEIN: And they themselves are in this endless process of evolution? There is no overall beginning?

DALAI LAMA: Exactly.

ANTON ZEILINGER: Where does one of these billionfold world systems emerge from?

DALAI LAMA: Space particles.

GEORGE GREENSTEIN: So it's not the universe that comes out of space particles, but the galaxies.

DALAI LAMA: Space particles can also be seen as the remnants of previous galactic systems. When Buddhists use the term *universe*, they are not referring to any particular galaxy system but to the infinite totality. The Tibetan technical term for universe means "that which goes through change and transformation," or "a reality that is subject to dissolution." That is the etymology of the term.

ANTON ZEILINGER: From what I just heard from Your Holiness about billionfold world systems, it seems as if Buddhism has known for a long time that stars are just other suns. Is that right? Did they know that before the West discovered it? It was a big discovery in the West.

DALAI LAMA: When you talk about Buddhist cosmology, you have to take into account two quite different, but not necessarily distinct, discourses. One is the Abhidharma system of cosmology, in which our galaxy is described. It also gives very exact measurements of the distance from the earth to the sun and moon and the stars, as well as the size of the sun and moon. The problem is, these measurements are wrong from the modern scientific point of view. For example, the sun is only bigger than the moon by a tiny fraction, and they are the same distance from the earth. These measurements are just crazy. The writer of this fifth-century text didn't have any telescopes, of course, but he probably also had very blurred vision! [*This remark brought forth laughter on all sides.*]

GEORGE GREENSTEIN: Were these distances determined philosophically or by some observation?

DALAI LAMA: Vasubhandu was probably drawing on the consensual view among astronomers and astrologers of his time. The point is, the Buddha himself, as well as his later followers, did not give priority to mapping the physical universe. They would only do that marginally or peripherally, and generally when they did so, they would do it in accordance with the views that were current at the time. Their priority was to understand the nature of the truth of suffering, the source of suffering, the cessation of suffering, and the path to the cessation—the Four Noble Truths. That is what they were really concerned about. In this regard, they were also centrally concerned with the reality of emptiness. Comprehending the reality of emptiness transforms the mind such that one's previous mistaken views are banished. We consider ignorance, the wrong conception of ultimate ignorance, to be a source of suffering and of wrongdoing. In order to change that, we have to develop the right view of emptiness.

Sentient Life Elsewhere in the Universe

Later in the week, there was a brief exchange about world systems and the presence of sentient life throughout the universe. I have interpolated it into the discussion here because of its relevance.

ANTON ZEILINGER: Your Holiness, when we talked about thousands and thousands of world systems, Alan mentioned that what you count are living world systems. Is that right? Do Buddhists believe that there are really living systems out there?

DALAI LAMA: Oh, yes.

ANTON ZEILINGER: And there are many, many of them?

ALAN WALLACE: When they speak of a billionfold world system, they don't count world systems that are uninhabited by sentient life forms. Only those that have sentient beings are even counted.

ANTON ZEILINGER: And how are sentient life forms defined?

DALAI LAMA: As an example, they include various animal forms and human forms through the process of evolution. And, of course, some can have different types of bodies made out of different substances. According to the Buddhist definition, a sentient being is a living organism that has the capacity for pain and pleasure. But in one of the previous Mind and Life Conferences, we had a long discussion that led to the general consensus that, at least in the context of life on earth, a sentient being could be defined as a living organism that can move by its own power. Even a single germ that can move is sentient.

GEORGE GREENSTEIN: How about a tree?

DALAI LAMA: A tree is alive but not sentient.

ARTHUR ZAJONC: Are sentient beings always substantial, material beings?

DALAI LAMA: According to Eastern thought, shared by both Hindus and Buddhists, in addition to physical beings that live within the so-called desire realm or sensual realm, there is also a whole range of sentient beings living in a subtle form realm, which is not grossly physical. They are born, they live, they die, but they don't have gross physical bodies. Beyond that there is also a formless realm, and there are likewise even formless sentient beings that are born, live, and die but don't have physical form at all. Moreover, there is an interpenetration of these realms. Even in the physical world here on planet Earth, there are said to be beings of the form realm, and quite conceivably the formless realm as well. Within Eastern thought, these three dimensions of existence are widely accepted: the sensual realm that we can see, the form realm, and the formless realm.

Of course, the paradigm of the fascination with extraterrestrial life is an image of weird extraterrestrial beings coming into contact with human beings on Earth. But from the Buddhist point of view, given the understanding that there are sentient beings in different realms of existence even within this world system, there are probably more explanations that could account for phenomena which otherwise would be considered mysterious. Even though there's no totally compelling scientific evidence of aliens coming to planet Earth, there are many individuals who have had very anomalous experiences, people who seem in every other way sane. There doesn't seem any way to account for this within the standard scientific paradigm, but the Buddhists might have an easier time taking these people seriously. There are all kinds of possibilities.

My most vivid impression from the day's work was the relentless application of logic to the questions of space, time, and the evolution of the universe by the Dalai Lama and the tradition he exemplifies. Experimental physicists like Anton Zeilinger and me are characteristically reluctant to speculate beyond what the data show. Yet quite often, especially in cosmology, the relevant observations are lacking or are even impossible to make. The methods of theoretical astrophysics, therefore, are reminiscent of those used by the Buddhists, in that both place great reliance on careful and consistent analysis, which often leads far beyond what is apparent in observation. Nowhere is this similarity more striking than in the story of the early universe now widely accepted within the astrophysical community.

I could not help noticing the correspondence between the view of the Dalai Lama and, for example, the "eternal, fractal, inflationary universe(s)" being advanced by contemporary astrophysicists like Alan Guth at MIT and Andrei Linde at Stanford. The Dalai Lama calls for a limitless universe within which a "billionfold world system" evolves. That is, he imagines many distinct parallel cosmoses like our own, each evolving according to its own laws and design. The inflationary scenarios currently being advanced in astrophysics also posit an infinite number of pocket universes that explode from fluctuations within the quantum vacuum into myriad and diverse cosmoses. These distant cousins of our cosmos are beyond our ability to observe directly. However, astrophysicists have produced many indirect, observationally based arguments that support the inflationary story of the very early universe, such as those utilizing the data provided by the Cosmic Background Explorer satellite (COBE).

Yet, far more important than any superficial similarity that might exist between Buddhist and Western scenarios of the early universe, the kinds of questions asked and their common confidence in thinking as a means to insight are identical in both traditions. Theorist Andrei Linde once asked in an article on the inflationary model, "The first, and main, problem is the very existence of the big bang. One may wonder, What came before? If space-time did not exist then, how could everything appear from nothing?

What arose first: the universe or the laws determining its evolution?"[5] How like the Dalai Lama's questions to us. Astrophysicists make use of giant telescopes, particle physics, and Einstein's theory of general relativity to develop their answers. The foundations of Buddhist cosmology lie in the remarkable power of careful thinking and close observation of both the outer and inner worlds.

5

Quantum Logic Meets Buddhist Logic

The rigorous mathematical foundations of modern physical science are familiar to everyone. A comparable dedication to logic and philosophical rigor goes far back in the history of Buddhism in India. The two great logicians of Buddhism were Dignaga (fifth century) and Dharmakirti (eighth century), who each wrote treatises on logic, the rules of syllogistic reasoning, and the modes of analysis that lead to valid conclusions. They are still studied today in the same way we study Plato and Aristotle. For centuries Buddhist monks have trained in classes, as well as through formal debate, on these texts and their commentaries. During the period of Dignaga and Dharmakirti, royal patronage (and therefore the existence of entire monasteries) depended on the outcome of public debates with Hindu Brahmins. Debate remains a central component of monastic training to this day and can involve decades of practice, using increasingly refined methods and more difficult subject matter.

Of the many streams within Tibetan Buddhism, the Gelugpa tradition, in which the Dalai Lama was trained, is the one most closely associated with these philosophical methods. In his autobiography, the Gelugpa monk-scholar Geshe Rabten put it this way:

> *Logic is studied to train the mind in subtle reasoning, thus enabling one later to appreciate the great scriptures. After developing his intelligence and discriminatory powers in this way, a monk is able to apply as many as twenty or thirty logical approaches to each major*

point of teaching. Like monkeys that can swing freely through the trees in a dense forest, our minds must be very supple easily to comprehend the depth of the concepts presented in scriptures.[1]

During the course of the second day, the importance of careful reasoning and logical analysis for both quantum physics and cosmology became evident. The Dalai Lama had raised questions about the kind of logic that governs quantum phenomena: Is it the same or different from the logic of Aristotle or Dharmakirti? In particular, is the law of the excluded middle still valid in quantum mechanics? The subject was again discussed on the final day. I have brought the two discussions together for greater coherence.

ARTHUR ZAJONC: During the tea break, some of us have had a discussion concerning monastic training in Buddhist logic and have compared it with the forms of logic used in classical physics and quantum mechanics. Can we pick up this topic?

DALAI LAMA: In Buddhist logic, one commonly speaks of a phenomenon being of one nature but having different terms. For example, Anton may be regarded as a certain woman's son, and he is a certain child's father. That certain child's father and that certain woman's son have the same nature. But there are two approaches, with different terminologies for the same entity, depending on context. Is there a similar structure or classification within physics? On that basis also, one speaks of mutual exclusivity of two phenomena having their own distinct natures. If something is A, it cannot possibly be B. If it is B, it cannot possibly be A. There is no third option. This runs throughout Buddhist logic and epistemology. In a previous Mind and Life conference, a question was raised as to whether the law of the excluded middle is tenable. At that time, the scientists responded that it was not actually tenable any longer. The grounds given were very different from the ones given today. The discussion so far this morning has not made it clear that the law of the excluded middle is untenable or even that it is being challenged.

Quantum Superposition and the Excluded Middle

ARTHUR ZAJONC: I wonder if I can make a simple first step toward addressing the question of the law of the excluded middle. Here is the calcite crystal we used this morning and two glasses. In classical thinking, we would say, for example, that I can put the crystal into this glass or I can place the crystal in the other glass. There is no way for me to place the crystal, without breaking it, into both glasses simultaneously. It would seem to be a logical impossibility.

However, if I am not working with a crystal but with an electron—a quantum particle—not only can I put the electron into something completely analogous to this glass, or into the second glass, but I also have a new possibility. I can put the electron into an ambiguous state that is nonclassical and is very difficult to think about. It is what we call a superposition state, where you could say the single object is in both places, in some sense that is hard to describe in normal language. It is an experimental fact that a single electron—not the electron in halves—is, in some sense, in both glasses. It's not merely a way of speaking or an arbitrary convention because there are specific experimental consequences of that particular, ambiguous state we call a superposition state.

DALAI LAMA: Is the electron in each place at a single instant or simultaneously?

ANTON ZEILINGER: Maybe one should clarify a little bit. The statement is not that the electron is in both places at the same time but that it is ambiguous as to whether it is here or there. The complete quantum description contains both possibilities. We are saying the electron is in a superposition of being in both places. We don't know definitely where it is, but the important point is that other new phenomena follow from this ambiguity. It's different from the situation where I could say the electron is definitely here or definitely there, but I just don't know which it is. That would be subjective ignorance. But with superposition, there is no way to tell whether it is here or there. We have to leave the ambiguity open, and then something new follows.

ARTHUR ZAJONC: New phenomena, and new experimental results, appear as a consequence.

ANTON ZEILINGER: And one of them is the interference pattern Your Holiness saw yesterday.

DALAI LAMA: When you see this superposition, it's not simply that it might possibly be in either, is it?

ARTHUR ZAJONC: That it could be either in this or in that, but only we don't know? That is not what we're talking about.

DALAI LAMA: Then what is it?

ARTHUR ZAJONC: This is the impossible description.

DALAI LAMA: So far all I am hearing is that it's ineffable!

Once again laughter broke out. However, in fact, many of the founders of quantum mechanics did believe that superposition, the central mystery of quantum mechanics, is ineffable. They felt that language was developed on the basis of normal sense experience, and quantum phenomena go far beyond what we encounter in the macroscopic world of the senses. Therefore, they argued, it would be forever impossible to reexpress the formal mathematical treatment of superposition in conventional language.

Hoping to make clear the new ambiguity of quantum mechanics, David returned to his example of polarization. Holding up his two polarizing sheets, he contrasted the case in which the two sheets are perpendicular to one another with the case in which they are oblique, which is another instance of superposition. David had made original contributions to the area of quantum logic and so was well prepared to clarify the ways in which the laws of logic have to be modified in light of quantum physics. In fact, quantum theory requires that we relinquish the distributive law, as will be demonstrated later, but David also brought into the discussion certain developments in pure mathematics. Kurt Goedel's incompleteness theorem, especially, has far-reaching consequences for formal logic.

Referring to the implications of quantum superposition, David began the discussion.

DAVID FINKELSTEIN: This is not a breakdown of the excluded middle, of which there have been some important studies. At one time it was thought to be a breakdown of excluded middle by the philosopher Reichenbach. This was very early in the days of quantum theory. Feynman quickly straightened it out and showed that this funny situation is a breakdown of the distributive law. The breakdown of the excluded middle and the breakdown of distribution both turn out to be consequences of incompleteness. In fact, these are the main changes in logic in this century, but the two have nothing to do with each other.

The Incompleteness of Logical Systems

DALAI LAMA: Are you suggesting that the nature of logic is such that it can never be complete?

DAVID FINKELSTEIN: Yes, for two different reasons. The most dramatic, perhaps, is the incompleteness suggested by Goedel that a system of logic rich enough to express arithmetic cannot be complete. In particular, it cannot answer the question of its own consistency. This is an incompleteness that seems to arise out of the problem of self-reference. It is impossible to know yourself completely in a formal system.

DALAI LAMA: There is an analogous problem in Buddhist epistemology:

Often the validation of the objective world is based on a validation of the cognition. Then the question arises of how to validate the cognition, or knowledge. As long as you cannot develop an epistemological system that includes a mutual validation or mutual dependence, then you need some other means of validation. Some Buddhist epistemologies began to postulate the involvement of an aperceptive faculty in cognitive events, such that any instance of cognition is accompanied by a faculty which is aperceptive.

DAVID FINKELSTEIN: There are propositions about numbers that can be believed without contradiction, but you can also deny them without contradiction. If by "true" one means provable, then they are neither true nor false. So one needs a logical system without the excluded middle. Such a system was developed by a Dutch mathematician, Brouwer.[2] It's called intuitionistic logic. It is very important in the theory of computation. Brouwer was deeply concerned about what it meant to say that a mathematical statement was true. He insisted that it ultimately meant that you could perceive it in your intuition clearly and completely. He was not content, for example, to prove something existed by showing that the denial of its existence led to a contradiction. That didn't tell you what the thing was.

DALAI LAMA: Is it true, then, that the theories of physics are coming closer to conventional language, whereas previous physics was quite abstracted or distanced from ordinary speech? My point is this: There has been a long-standing assumption that there is something absolute out there. You assume that is the case, and then you seek to understand its nature. As long as you think it is there, then you feel you can come to some definite truth: It is black, or it is white. It is A; it is not-B. But then you seek out its nature and you don't find it. Through minute and subtle analysis, you come to a point where the whole notion of reality begins to become problematic. Also, you realize that there is no clear separation, no division as we imagine in classical logic, between A and not-A. One may then come to a growing appreciation of the imputative quality of much of the usage of our language—the fact that when we talk about certain things, in some sense, we are participating in a language which has more to do with convention rather than direct reference to things.

DAVID FINKELSTEIN: I would not say convention so much as relative.

DALAI LAMA: In Buddhism it would be called dependent origination or interrelatedness.

DAVID FINKELSTEIN: I think that most of humanity lives in a world of flux. Three hundred years ago, physicists encountered a little rock in the river and they all climbed on it. It lasted for three hundred years. Now we have to take to the boats with the rest of humanity.

Universals and Indistinguishability

DAVID FINKELSTEIN: Also related to this is whether the notion of universals or some kind of generality is tenable—whether you can even speak of universals as opposed to specifics alone. For example, we talk about this specific pen as opposed to that specific pen. At the same time, we instantly recognize a commonality between the two, which suggests that we have some kind of concept of a universal pen. The question is whether that universal pen is tenable or merely a convenient category that we use. Of course, if we were to search for the universal pen that pervades both these two instances of pen, we would not find it. In the same way, if we were to search for a specific point, or whatever, in quantum physics, it remains untenable. At the same time, the existence of a referent of the term *pen* can be maintained, without any specificity, which is also the object of our perception and concept.

What is the understanding of universals in physics? Are such designations as electron *and* photon *merely convenient conceptual categories, or is there a real referent? As in India and Tibet, the debate among scientists and Western philosophers on this question has gone on for centuries under the name of nominalism (or relativism) versus realism. We are well aware of the phenomenal existence of objects like pens, but is there such a thing as a "pen in itself" that is distinct from the specific appearance of particular pens? On the one hand, many scientists (and certainly most philosophers of science) eschew a simple metaphysical realism that presumes a domain of reality somehow more fundamental than ours and which is beyond all subjective experience. And yet most scientists have a robust sense of physical reality and are convinced that they are not merely playing language games with socially constructed concepts. No, scientists by and large see themselves as discovering real truths about nature, not somehow inventing them. Most scientists are cautious realists, but are they mistaken? The dichotomy of nominalism versus realism is resolved in Tibetan Buddhism by the development of a third, or middle, way: Madhyamika philosophy. A comparable middle way is largely missing in the Western debate.*

However, I would like to sidestep this debate and make a more grounded set of remarks. Mature classical physics can, it seems to me, be understood within a purely nominalistic framework. In the laboratory, we always deal with specific masses, speeds, directions, and so on. When we invoke the abstract concepts of point masses, moments of inertia, and the like, we do not grant them an ontological status above and beyond the phenomena. We see them as convenient conceptual constructs. However, another category of concepts common to modern physics seems, in my opinion, to merit more attention as candidates for universals. Consider the concept of electron. The "indistinguishability" of particles like electrons is a bedrock principle of quantum physics. That is, there are absolutely no intrinsic features by which we can distinguish one electron from another. Of course, electrons can be in different states, but the electrons themselves are indistinguishable. Given two electrons, if I can switch one for the other, no experimental observation can detect the exchange. This fact has profound physical implications. The covalent bonds that hold molecules together are due to this exchange degeneracy. Likewise, all the mysteries of the EPR effect, which we treated earlier, rest on the principle of indistinguishability. Without this principle, our world would not exist as we know it.

Every pen in the world is distinct from every other pen in some way, large or small. But in quantum theory, every electron is exactly the same as every other. Should we not say, therefore, that this deep symmetry in nature reflects a universal feature recognized by physics? When it becomes impossible even in principle to maintain awareness of the particular, do we not then have grounds for granting standing to the universal, especially when the consequences are so dramatic? Unfortunately, we did not take up this topic directly, but we did learn something of the Buddhist critique of metaphysics.

DALAI LAMA: In Buddhism, there is a tension between two systems of epistemology. One arises out of a realist school, where the very criterion of a valid cognition is some truly existent entity. You determine whether a cognition is valid in relationship to some truly existent, absolute phenomenon. There is another system of epistemology—the Madhyamika view—where the underlying metaphysics rejects any possibility of intrinsic entities, any kind of absolute reality. In this system, the criterion for valid cognition does not refer to an absolute, truly existent phenomenon.

The question is, How do you distinguish valid and nonvalid cognition without some external, independent referent that is truly existent? Since you don't have any absolute referent, there can be no correspondence theory. There are two ways to approach it: If you have a cognition that is not invalidated by another valid cognition—by conventional valid knowledge—then you can say it stands. Or, if you find that the cognition is not negated by a very careful, critical mode of analysis that probes into the ultimate nature of that reality, then it stands.

Correspondence theory maintains that cognition is valid when that which is cognized stands in correspondence to a truly existent entity (an absolute referent) that remains unexperienced. In this way a correspondence between experience and reality is established. A lengthy conversation in Tibetan followed.

ALAN WALLACE: I said to His Holiness that this sounds problematic because, in the first instance, when a cognition is valid because it is not invalidated by another valid cognition, how do you know whether *that* cognition is valid or not? Any number of people may share the same misconception and support each other, but it is all false. How does that stand up?

Between Relativism and Realism

DALAI LAMA: Once you do away with any possibility of grounding epistemology in a truly existing external world, or internal world, for that matter, then the only option you have is to develop an epistemological system where there is a mutual dependence between subject and object. This is the basic approach of the Madhyamika system: that in some sense the reality of the object is validated by the cognition, and the cognition is validated by the reality of the object. You cannot really separate the two. They are so intertwined that to talk about a valid cognition without reference to the reality of the object is simply, one could say, nonsensical. And similarly to talk about the reality of an object without a verifying cognition is again nonsensical.

ALAN WALLACE: It's a bootstrap situation.

DALAI LAMA: There is a distinction made between not finding something and negating something. Although, in ultimate analysis, one may not find the object under investigation, that does not negate that very object. Here the Madhyamikas would turn for support to the notion of convention, or consensual agreement, which is problematic. There could be many different types of consensual agreement in relation to an object under investigation.

For instance, there is consensual agreement that this is a glass. It's not an ultimate statement, but we all agree that this is what we call a glass. Even if someone disputes that by saying that it is something else, that proposition can easily be invalidated by the perception of the object and the knowledge that, consensually, this is a glass. However, Alan has brought up the example of Nazi Germany, where there was a mass invalid cognition among many Germans in the 1930s that all of their economic problems were attributable to the Jews.

ALAN WALLACE: A lot of people supported each other in that view, but they were all wrong. It doesn't matter if there were 10,000 or 10 million. Reality is not decided by vote.

DALAI LAMA: In the case of value judgments, one draws on a different kind of consensual agreement. The statement that the economic problems of Germany were the fault of the Jews is not a value judgment. It's an invalid statement of reality rather than a statement of value. A pure value judgment would be to say that a particular type of art form is good. If a lot of people agree that impressionistic art is good art, no one can invalidate them. In other words, the validity of a value judgment can stand by consensual agreement alone. For that group of people it is a true statement because they think it so. It's not true that the Jews were the cause of the problems of Germany just because many people thought so. These are different types of situations.

The Origins and Development of Logic

On Friday we returned to the issue of logic. After his full monastic training in Tibetan Buddhism, Thubten Jinpa studied philosophy at Cambridge University in England and therefore had a nearly unique experience of both Buddhist and Western philosophical styles.

THUBTEN JINPA: At first glance we see quite a lot of difference between Buddhist logic and Western logic, especially in the criteria used to validate a logical argument. In the Western context, logic has only to do with form rather than content. It is, in some sense, completely divorced from epistemology, whereas in Buddhism the epistemological dimension enters into the validity of a logical argument. A tautological argument, which may be valid in the Western context, cannot be a valid argument in Buddhism. Still, the underlying logical principles are the same—the principles of noncontradiction, identity, and difference.

DALAI LAMA: I am trying to understand why Western logic and Buddhist logic seem to have developed in slightly different ways. Am I right in thinking that Western logic developed on the basis of analysis of the physical world? Can it be applied equally to manifest physical phenomena and also to the mental realm and to abstract composite entities?

Buddhist logic developed in such a way that it can be applied not only to the physical world but also equally to the mental world and the abstract composite entities.

ARTHUR ZAJONC: This is an interesting question, which perhaps we can explore further. What is the basis for logical inquiry? Is logic drawn from the world of experience, where we manipulate objects and the form and character of that world imprints itself on our mind, and therefore the order of analysis is determined by the order of the world? This is one view that has proponents in the West—I think Aristotle was one of them. But there are others who speak to an innate or genetic capacity for logic, whether it derives from our biological organization or some other basis, which is inherent in the human being independent of experience. Is there something comparable to these two views in the Buddhist tradition, or is there understood to be another basis for logic?

PIET HUT: A related question is whether the logic is universal that is discovered, for example, by manipulating the objects of this world. Or do the laws of logic differ for particular domains? Are the laws of logic that apply in the domain of classical objects identical to those found as one moves from the sense world and material existence to a more subtle domain of experience? There are suggestions—David would say facts, and I think most of us in this room would probably agree in some measure—that strictly applied classical logic is inadequate for understanding the full range of quantum phenomena. Are there parallels in Buddhist philosophy, where, as you extend experience, the logic of one domain is no longer operable or fully appropriate in a second domain?

DALAI LAMA: My own view of logic is basically the first, a posteriori position. I feel that logical forms and principles are very much derived from our experience of the sensory world. If you look at animals, nonhuman species have a quite limited need for rational thought or some kind of logical system, and they are capable of surviving in the world on the basis of the limited capacity they have. A cat might well be engaged in some form of thinking, like what is the best way to catch the mice, but animals deal mostly with manifest phenomena: the cat, the hole, the mouse. . . . They don't need to work through syllogisms and logic to meet their needs for survival, procreation, and so forth.

But human beings have much more sophisticated needs and also a much higher level of curiosity. There is a greater interest, as well as a greater need, and we are equipped with some ability to fulfill this desire. With curiosity you enter the realm of concealed phenomena. For that, you must have recourse to logic and inference. A variety of different modes of analysis are evoked by the different types of concealed phenomena you seek to understand: modes of analysis pertaining to function, to the nature of phenomena, to dependence. This pertains not only to our curiosity about the external physical world but equally to our inquiries about the nature of mental phenomena. In both cases you are dealing with evident phenomena that immediately present themselves to experience.

On the basis of this experience, one then applies modes of inquiry. For example, the question of whether two phenomena share any common ground or are mutually exclusive is not something you dream up in a vacuum. Rather, you observe manifest phenomena, and then you impose conceptually upon these manifest phenomena. However, that does not belie the fact that there are also other types of phenomena that are purely conceptual constructs.

THUBTEN JINPA (*commenting on a side discussion that took place in Tibetan between himself and His Holiness*): I was still trying to argue an a priori position, and I raised the point that Dharmakirti's text has a very memorable line where he says that all the parts of a syllogism—the subject, the predicate, the reasoning, the identity relationships, the difference relationships, the contradictions—all of these are nothing but constructs by logicians. Let's say that we were arguing that this glass is impermanent because it is breakable. So far as the empirical object is concerned, there is only one thing there, whatever you may say about the glass as the subject or impermanence as the predicate or breakable as the reason.

DALAI LAMA: But even in this case, you need an empirical object on which to build. I mentioned that different schools of Buddhism hold different fundamental assumptions about the nature of reality. If you take as a starting premise that phenomena are self-defining and have their own intrinsic nature, then you develop a certain logic based upon that. In this case, as soon as you are studying valid cognition, or *pramana*, you are studying logic. The very notion of logical reasoning is inseparable from inference, which is inseparable from the criteria that determine what is and what is not valid cognition. So logic and epistemology are unified in the Buddhist education that is based upon this realist assumption.

On the other hand, in the Madhyamika school in particular, there is nothing that bears its own intrinsic existence uniquely from the side of the object, which implies that all appearances are illusory in a very significant way. When you say that the nature of appearance is illusory, this suggests a fundamental disparity between sensory appearances and the actual mode of existence of the phenomena that are appearing. Illusory means there's a lie here: There is something misleading in the very nature of appearances themselves because they belie the nature of the existence of the phenomenon that is appearing to the senses. But even if you take that as your starting premise, it is still imperative to have some criteria for determining what is and what is not valid cognition. You don't throw judgment to the winds and say whatever you like is true. Now, as you establish the criteria for valid cognition, a crucial form of valid cognition is inference, which means once again you go back to logic. So you must have a kind of logic that is still relevant even to appearances that are illusory.

Because of this fundamental difference in the metaphysical assumptions of the two logical systems, there is an important debate as to whether a meaningful object could be established by means of common criteria shared between disputants from the two schools. The nonessentialist would argue that this is entirely meaningless. Logical argumentation takes place only on the basis of consensual understanding rather than mutually verifiable criteria of validation. For the Madhyamika, the issue of verbal and conceptual consensus is crucial. For the realist, there is no need for consensus; one only need look at the nature of reality itself.

Another point is that within the Madhyamika system there are two basic modes of analysis: ultimate and relative, and they pertain to two different domains of reality. Ultimate analysis is not satisfied with mere appearances but probes into the fundamental nature of the phenomenon in question and leads to the ultimate truth of emptiness. Relative, or conventional, analysis is satisfied with mere appearances and so works within the context of these illusory appearances. Most of science would fit into that category. The investigation of organisms, cells, neurons, and so forth is all relative analysis. Science doesn't try to probe the ultimate nature of neurons.

Buddhist logic includes various types of inference, together with reasons that support and give rise to the inference. For example, inference may be based upon an affirmative syllogism: This is so because that is so. The affirmative statement implies an affirmative conclusion. Within affirmative reasoning, you also have different types of inference. You may infer the cause on the basis of the result, as in causal relationships. The typical example given is that we infer the existence of a fire because we can see smoke. (Perhaps this example has something to do with the fact that Tibetans on the whole are nomads, traveling from place to place, for whom the perception of fire is very important.) Inferences can also be made on the basis that two phenomena are of the same nature.

Technically, this is called same nature inference, and a typical example is the one given earlier, that this glass is impermanent because it is breakable. The impermanence and breakableness of the glass are of the same nature as the glass itself: There is no causality or sequence. Alternatively, you can use negative reasoning, a syllogism that negates a certain proposed entity and demonstrates that it does not exist. For inference through negation, we turn around the syllogism of causal analysis and say, for example, if a lake has mist at night, which might be construed as smoke, you can infer the absence of smoke because there cannot be fire in the middle of a body of water.

All such reasoning always refers to some object of experience. There is an affirmative object that you are dealing with, or there is an absence that you can actually perceive. For example, we can perceive right now the absence of an elephant on the table. We can reason about that absence because causality takes place in the natural world. All of these different types of reasoning are always based upon something in experience and are not created in isolation.

I would be willing to move a little closer to Thubten Jinpa's a priori position in regard to two modes of reasoning. One is syllogistic reasoning, where you claim this is so because that is so. It's a little bit pushy, trying to compel somebody to think your way: This glass is impermanent because it's breakable. The other mode is consequential reasoning, where you listen to the other person's position and then lead him from there to the consequences that follow from his position. Using the same example of the impermanence of the glass, if you were to assert that it is permanent, I would take this to mean, in consequence, that it's not breakable. So, is the difference between these two modes of reasoning, syllogistic and consequential, purely experiential? Do we find different types of experience or different phenomena that elicit these two modes of reasoning? I don't think so. They seem to be more like different strategies for dealing with the same experience. In other words, they really do seem to be more like a priori conceptual constructs that are imposed upon experience, rather than a mode of reasoning that is elicited by experience.

DAVID FINKELSTEIN: When you brought up Dharmakirti's statement about the constructs of logicians, it suggested an answer to the question that you rose earlier concerning the difference between Western and Buddhist logic. When Newton tried to project his understanding down to the very small, he simply took his system of the planets and shrunk it. Everything, he said, is made of very small, hard things, which bounce off each other, and each of them lasts forever. There's a lot of truth in

Newton's model, even today. Laws of conservation of energy and momentum work in the macroscopic world. I'm trying now to remember something I read over thirty years ago, but I recall that Dharmakirti really addressed something else in the macroscopic world: the doctrine of karma. Perhaps he took seriously Aristotle's declaration that everything in the world that changes has what he called an entelechy or a soul or a monad. Entelechy, he said, is that which converts the potential into the actual.

DALAI LAMA: This is probably similar to karma.

DAVID FINKELSTEIN: Dharmakirti said that the things in the microscopic world are born, originate, propagate, meet other things, annihilate mutually, and then dependently reoriginate. That captures a remarkable amount of the truth about what goes on in the microscopic world. It's an interesting question, why he was so lucky. At any rate, it indicates a big difference between the Western and Buddhist approaches. In one case, you start from mechanical truth, and in the other case, you start from spiritual experience. I think the same is true of the logics, also. There's a very simple iron-clad logic that even dogs know. It's not true that you have to be a person to reason logically. A famous philosopher—I'm not sure if it was Hume or Kant—described what he claimed to have seen: A dog was chasing a rabbit along a trail. The trail forked and the dog did not see which way the rabbit went. The dog followed one branch of the trail and realized the scent of the rabbit had disappeared. But instead of retracing his trail, the dog took a shortcut over to the other trail. It's irresistible to say that the dog has reasoned: either A or B; not-B; therefore A. If he didn't reason that way, he would not have had his lunch. It's not that logic does not follow from experience, but it could follow from experience many generations back. It could be born into us as part of our instinctual apparatus. There probably *is* a natural logic, and part of the problem with quantum theory is that it seems unnatural. Photons do not behave like rabbits.

Novel Features of Quantum Logic

At this juncture I saw an opportunity to clarify a point that had been confusing the Dalai Lama and has confused many others when they first consider quantum mechanics. In exactly what way is the law of the excluded middle retained in quantum logic, and what is the law of logic that is violated? Nearly everyone around the table took part in the following discussion. David Finkelstein and I explained aspects of quantum logic, and An-

ton Zeilinger drew diagrams on overhead transparencies, illustrating precisely how the law of the excluded middle remains valid while the distributive law becomes invalid in the quantum domain. I began by illustrating the law of the excluded middle with a glass and a pen.

ARTHUR ZAJONC: May I use this as a place to add something important, Your Holiness? You have asked about the law of the excluded middle on a number of occasions. I want to make sure that when you leave this afternoon that you are not misinformed.

If I have only one glass and one pen, then I have two possibilities. The pen is in the glass (call this P), or it's not in the glass (not-P). The pen is either in the cup or it's not in the cup. This is an example of the law of the excluded middle.

DALAI LAMA: From the Buddhist point of view, there's a slight problem if one were to say that the pen is either in the glass or on the table. It has to be either the pen is in the glass or not, even though we can see that the pen is on the table.

ARTHUR ZAJONC: Yes, I agree. To be precise we must say the pen is either in the glass or not in the glass.

ALAN WALLACE: Then they are diametrically opposed.

ARTHUR ZAJONC: Yes. Now consider the double-slit experiment that Anton performed at the beginning of our meeting. When you first meet this experiment, it appears as if this law is violated. To make the situation clearer I'm going to use a written notation, which is intended to represent the state of the system. In the first illustration [*figure 5.1*], I show the two options schematically: the pen inside or outside the glass. We have one state (pen inside the glass) and its negation (pen outside).

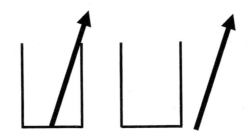

Pen in the glass   Pen not in the glass

Figure 5.1 The law of the excluded middle states that the pen is either in the glass or not in the glass.

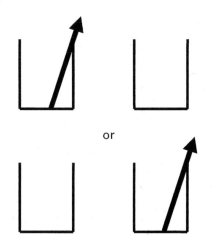

Figure 5.2 These are the two logical possibilities for a classical particle that is traveling through double slits. The top represents a particle through the left slit, the bottom through the right slit.

If you go to the quantum level, you can ask, is it possible to describe the state of a photon in a similar quantum language? In order to do so we must introduce, as we have seen, the concept of superposition. Your Holiness saw that in classical language, we say a photon goes through one slit or the photon passes through the other slit. We can represent these two options in a manner analogous to the glass/pen example.

I sketched the double-slit experiment and represented the two possibilities with a pair of water glasses. When the left glass contains a pen and the right glass is empty, this represents the classical trajectory of the photon through the left slit. The image is reversed for the other trajectory (see figure 5.2).

Quantum mechanically, we have another and different option, namely, superposition, and we indicate this with a plus (+) between the two classical options. This is a definite quantum state. You can ask, is there a negation of this state, and the answer in quantum mechanics is yes. There is a simple negation of this state; the only difference is the minus (–) sign. If the quantum superposition state with the + sign is P, then the one with the – sign is a different superposition state we can call not-P. [*See figure 5.3.*]

ALAN WALLACE: What does the latter mean?

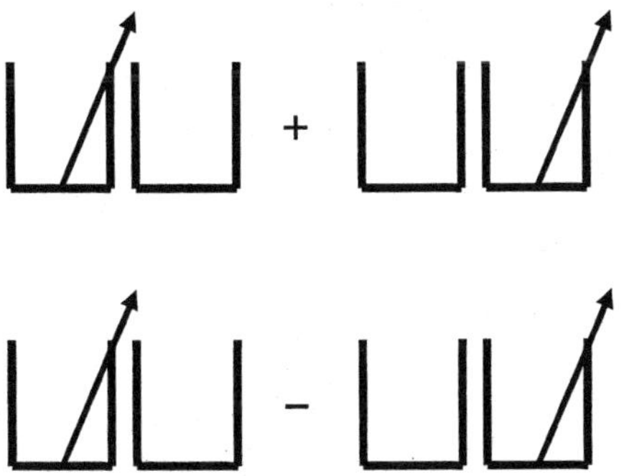

Figure 5.3 Superposition states of the double-slit experiment. The pen in the glass on the left represents the trajectory in which the particle goes through the left slit. When the pen is in the glass on the right, the particle travels through the right slit. Two superposition states exist, as shown in the figure.

ARTHUR ZAJONC: We could use circular polarization—right and left circular polarized light—to demonstrate the meaning of this state.

This provoked a gale of laughter. Ever since the complex use of polarization in David's presentation, we were all clear that we were better off using other illustrative examples. Anton jumped in with his familiar double-slit experiment.

ANTON ZEILINGER: I know a simpler example. If the upper statement with the plus describes the interference experiment I showed Your Holiness, with its bright stripes and the dark stripes, then the lower statement with the minus would describe the same experiment, but with the dark and the bright stripes reversed. This is the picture you get on the screen if you use the quantum state with the plus sign (+). [*Anton quickly sketched a set of alternating light and dark bands; see figure 5.4.*] If you use the one with the minus sign (–), you get the complementary picture. Where it was dark before, it is now bright, and where it was bright before, it is now dark.

ARTHUR ZAJONC: If the first case is P, then the second is not-P. So, formally, quantum mechanics does *not* violate the law of the excluded middle.

DALAI LAMA: This is a relief.

This brought a gale of laughter, especially from the Dalai Lama.

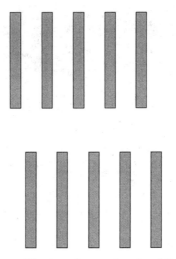

Figure 5.4 The interference bands of light and dark that are caused by the two types of superposition states, with (+) and (-), are shifted with respect to one another.

ARTHUR ZAJONC: But there is a law of logic that is violated. It's a different one. Let David talk about it, using the same double-slit example.

DAVID FINKELSTEIN: In this experiment, first the photon goes through a very small hole. This is an important part of the experiment—to make a coherent source.

ARTHUR ZAJONC: Should we draw a picture as you talk?

Anton got up and began drawing as David spoke (see figure 5.5).

DAVID FINKELSTEIN: Now draw a diaphragm with a little slit, one slit, and then a little farther away, a diaphragm with two slits. Call the first slit A. Call one of the second slits B and the other one not-B. In this experiment, when you see the electron hit the screen, you can say A and B or not-B. That is true. But you cannot say A and B. I will explain why in a moment. And you cannot say A and not-B. They are both, in fact, false.

ANTON ZEILINGER: So, we cannot say, A and B or A and not-B.

DAVID FINKELSTEIN: Correct, both are false.

It helps to use the diagram as a reference. We can correctly assert that a photon that has arrived at the screen started at the small hole A and went through either B or not-B. Oddly, from the standpoint of quantum me-

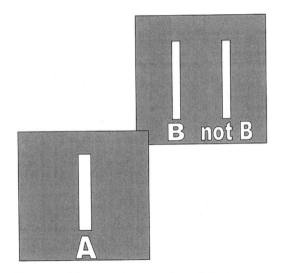

Figure 5.5 Electrons that pass through A also pass through B or not-B.

chanics, this is not the same as asserting that the photon began at A and went through the hole labeled B, or the photon began at A and went through the other hole labeled not-B. Anton wrote the two statements, using the formal notation of symbolic logic:

1. A ∧ (B ∨ ¬B)
2. (A ∧ B) ∨ (A ∧ ¬B)

In classical logic, the law of distribution holds and the two statements are equivalent. In quantum logic, they are not equivalent.

DAVID FINKELSTEIN: In classical logic, the two logical statements are the same. In quantum logic, the two are related by an inequality. If the top one is true, the second one is true. But if the bottom one is true, the top one can be false. To repeat, what is known is that a photon was emitted at A and reached the screen. You can say, A and (B or not-B).

ANTON ZEILINGER: So, you're saying that it must have gone through the middle screen somewhere.

DAVID FINKELSTEIN: And there are many operations you can perform to check this. But you can also show that "A and B" is false and that A and not-B separately is false.

"A and B" means that we know the photon travels from A to the screen through slit B. If we know which slit the photon passes through, be it B or not-B, then the interference pattern disappears, thereby falsifying the result. Knowledge of which way the photon traveled is sufficient to destroy the interference pattern. The ambiguity (B or not-B) is essential for this archetypal quantum phenomenon to appear.

ANTON ZEILINGER: Which means in simpler words, you cannot say that the photon either went from A through B or it went from A through the other one. This you cannot say.

DAVID FINKELSTEIN: This analysis hinges on giving up intuition. We're no longer dogs following rabbits. We recognize that logical words like *and* and *or* are constructs and that we give operations for verifying, for defining A and B; and when you know A and you know B, you can specify operations or actions for verification. In this case it works out that "A and B" is false in just the same way as when we have two polarizers at forty-five degrees. We say false even though some photons get through because there is no way to make photons that surely get through.

Laughter, once again, arose around the room at the use of polarization to explain the logic of quantum physics. There followed a lively exchange in Tibetan, leading to nods all the way around. It seemed that we had finally succeeding in clarifying an oft-misunderstood point in quantum logic. I asked if the Dalai Lama would like tea now or not. In keeping with our previous conversation, Piet Hut quipped, "Tea or not Tea?"

6

Participation and Personal Knowledge

During the afternoon of the second day, Tu Weiming made a short presentation of the new relationship between knower and known that modern physics seems to portend. In his words, we are in a "joint venture," a participatory relationship with the observed. As opposed to a distant and depersonalized knowledge, much in modern science and modern scholarship appears to argue for a personalized or embodied form of knowing, which remains, nonetheless, objective. The intimate investigation of human consciousness common to Buddhist and other forms of spiritual practice would seem to lead us in a similar direction. What might be the form of a fruitful relationship between these two modes of investigation and knowing, one (scientific) leading to the outer world and the other (spiritual) leading to the interior domain of consciousness?

TU WEIMING: We are being treated to very rich and subtle food for thought. I would like to focus on one idea that has emerged in the discussion: the idea of the observer—the person who knows. One of the messages of Anton's presentation was that the information obtained from these very refined experiments is a joint venture between the observer and the world out there, through the mediation of the instruments. The observer and the world out there enter into a joint venture and together generate a kind of information. I suspect that Anton feels that information itself may be the arena in which we will further explore the full implications of this joint venture. David also pointed out earlier that, as relativism is generally accepted, the complexity introduced by the observer becomes obvious and significant. In both cases the knower or the observer becomes vitally important.

With this as the central issue, I would like to share a thought about two kinds of enlightenment: the European and the Asian, particularly the Buddhist. Within the European enlightenment, there are also variations. The French enlightenment emphasizes the revolutionary spirit and individualism against clerical religion, whereas the Scottish enlightenment puts more emphasis on skepticism, empiricism, and pragmatism. Among the great values that came out of the European enlightenment, liberty, equality, individualism, human rights, due process of law, and rationality featured most prominently—especially instrumental rationality, or rationality that helps us to obtain knowledge in pragmatic and instrumental ways. The seventeenth-century philosopher Francis Bacon defined knowledge as power, different from the Greek idea of knowledge as wisdom.

The image held by the European enlightenment was one of rationality as a source of light; as knowledge was extended throughout the world, ignorance would gradually disappear. If we continued to shine the light of rationality on the universe, just like shining light into a room, all of the furniture and dimensions of the room would become known. That optimism obviously is no longer tenable, based upon the development of new physics and new sciences. The faith now is untenable that was dominant in the classical enlightenment—faith in the knower's ability to extend knowledge itself. The assumption now is that, as our knowledge horizon extends, our need to know more becomes more compelling and, in a way, ignorance extends further. This is special to the human condition. It's a cognitive deficit and an affective surplus. Our ability to know always falls short of our emotion, our intention, our need to know more. There is always an overflow of emotion, and there is always the deficit of being able to truly understand.

The primary concern of virtually all spiritual traditions, Eastern and Western, and in particular the Buddhist tradition, is self-knowledge: understanding and realization of the self through the science of spiritual exercises and the art of self-cultivation. In this sense, philosophy is a way of life, consisting of spiritual exercises to explore the inner landscape. Your Holiness mentioned the neglect of the inner landscape of the human condition. The focus of the enlightenment of science is to explore nature, the outer landscape.

The emergence of the new physics as a method strongly suggests that the exclusive dichotomies featured prominently in Descartes and many other enlightenment thinkers—the body-mind, spirit-matter, subject-object dichotomies—would have to be replaced. This either-or thinking would have to be replaced by a much more fruitful, nuanced, and layered integration of surface and depth, inside and outside, the part and the whole, the root and the branch. It is this ability to appreciate fruitful ambiguities, rather than to search for that which is true and certain in a limited sense, that opens up all kinds of new possibilities.

Self-Cultivation and Enlightenment

Weiming's phrase "fruitful ambiguity" captures beautifully the complex notion of superposition we have been laboring to illustrate. The ambiguity about the path of the electron is not a lack or failing, quite the contrary: It is a positive ambiguity that allows for an entirely new domain of phenomena and technical development.

DALAI LAMA: By ambiguity do you mean simply a lack of clarity, or do you mean something more powerful?

TU WEIMING: Certainly not a simple lack of clarity because it is fruitful, with all kinds of potentiality and possibility, an inexhaustible supply of things that we do not know yet.

This integrated vision of the knower demands a beneficial interaction between the two kinds of enlightenment—the Western enlightenment, focusing on instrumental rationality, exploring the external landscape, and the Asian, particularly the Buddhist, enlightenment. Many of the terms and ideas that were considered suspect before—dialogue, communication, mutual interaction, intersubjectivity, relationality, interconnectedness—all are absolutely critical for the training of the new knower. The knower is no longer simply a limited notion of the rational animal. In context, the knower would need to mobilize not only mental resources but also bodily resources, not simply the cognitive functions of the mind but also the affective dimensions of the heart and even the body.

It is in this sense that the concept of embodied thinking, or what the scholar Michael Polanyi termed "personal knowledge," becomes very, very important. Personal knowledge is not subjectivistic. It is not private. Profoundly felt or even embodied, personal knowledge can be publicly accountable. It can be debated and argued interpersonally. I think the time is ripe for imagining a new kind of education. It is highly desirable, maybe even necessary, that this new education integrates the self-cultivation of the Buddhist and other traditions. The establishment of these spiritual disciplines in the training of the modern academy, including the highly specialized natural sciences, is important because it will enhance one's self-awareness as an observer, not simply individually. It will enhance the communal, critical self-awareness of some of the most creative and reflective members of the scientific community. This is absolutely necessary for a new breakthrough.

The time is ripe for scientists—not just physicists but scientists in general—to appreciate the insights of Buddhist enlightenment as a form of self-cultivation. If I could ask Your Holiness personally, as a spiritual master who is amazingly open to modern science (you are probably the busiest person among us, and yet you have a spiritual quality of openness): How might a modern thinking person seasoned in highly complex, specialized knowledge benefit from Buddhist insight?

DALAI LAMA: It is difficult to say something offhand. I appreciate the atmosphere of warmth here, which is reflected in the smiles on the faces of all the participants. It is my fundamental belief that our basic human nature is gentleness and affection. Of course, for some individuals there is more stress on knowledge, but still the basic human nature is there. For a scientist who spends twenty-four hours thinking only about analysis, even a few seconds daily spent on cultivating compassion would be helpful when passing through a difficult period or painful experience. Even this would help to ensure that there's some natural response of compassion and concern. Obviously, I also think our knowledge should not be used for destruction. That is very clear. One may be curious and happy to gain more knowledge, but certainly it should not create pain or troubles for human society, for a group of people, or for individuals.

The two interpreters consulted and then asked Weiming whether he meant to include the Buddha's state of enlightenment in his considerations or only the more general discipline of self-cultivation.

TU WEIMING: Oh, yes, the enlightened state as well. The question that really challenges me is the notion of spiritual fruition, which leads to the idea of all 3,000 worlds realized in one instance. But that vision, based upon compassion and insight, is certainly the vision of an accomplished spiritual meditator. It is not subjective alone because it is communicable and it can help to transform the world. How can people like ourselves, and scientists in other areas, connect with that particular experience as a way of not only understanding ourselves but also doing our own work in many different specialized fields?

ARTHUR ZAJONC: Could I rephrase it in a nutshell? There is a knowledge of the external world, which we've discussed. There is a knowledge of the inner world. The two kinds of enlightenment point in different directions: the European enlightenment to outer accomplishment and knowledge, and the Buddhist enlightenment to inner accomplishment and knowledge. What is the relationship between these? Are they isolated and separated? Or is there some way of bringing the two into a fruitful relationship?

DALAI LAMA: There is definitely a relationship. Obviously, we need external development, as well as internal development. I feel the external development is a condition of our inner satisfaction. Generally speaking, the purpose of exploring reality, aside from fulfilling our inner satisfaction, is to benefit humanity. Up to now the model for science seems only involved in phenomena that can be measured or calculated. At the same time, twenty-four hours a day, we all experience another very important phenomenon, which is the experience of feeling. It is one of the most important factors that relates to our experience of joy and pain.

Many seemingly deep matters are only conditions that support increasing satisfaction and minimizing pain. There are other ways and means to increase satisfaction and happiness and to reduce pain, which are equally important. Whether one believes in religion or not, I think those ways and means are very important. So, while we are exploring the external world, it is equally important to investigate our minds and our mental function, particularly in the area of emotion. What kind of emotion is beneficial? What kind of emotion is destructive? How can we increase positive emotion? How can we reduce negative emotion?[1]

These brief remarks by the Dalai Lama were only an initial response to the large and important question posed by Weiming. Weiming had sought to invert the commonplace stereotype of subjective feelings as standing in opposition to purely objective thinking. He shifts us toward personal knowledge and objectified feelings. His remarks make an excellent transition to the considerations taken up in day three, which was my day to present. I had chosen to explore the relationship between experience and theory in physics in the hopes of finding a new grounding not only for scientific knowledge but for aesthetic and moral knowledge as well. In my view this requires the kind of reorientation Weiming is hinting at.

7

The Relation between Scientific Knowledge and Human Experience

Our attitude toward Western science has been deeply influenced by its outer successes. We grant its statements great authority, even when these are very distant from our personal experience or the direct experience of anyone. The power of reason has led us to infer the existence of elementary particles with unheard of properties. For example, the electron is thought to be a point particle, having no extension and yet possessing mass; and photons are quantized forms of pure energy that have no rest mass, and yet their trajectories bend in a gravitational field. Through examples, I traced the development of science from the study of the dynamics of large bodies to research into the more elusive nature of color, electricity, and the atomic structure of matter. As a civilization, we have come to grant fundamental status to the primary entities of physics and have come to view sense experience as an adaptive strategy (in the neo-Darwinian sense), with only a tenuous connection to the true nature of the world. Years ago I had become interested in a phenomenological approach to science, such as that advanced by the German poet Goethe or the philosophers Edmund Husserl and William James. I was interested to see how Buddhist philosophy handled similar issues.

ARTHUR ZAJONC: Today I would like to continue our discussion of quantum mechanics and also to discuss briefly Einstein's relativity theory. In doing so, I would like to connect these discussions to human experience and to ask what the relationship is between experience and theory.

What is the relationship between our everyday life experience—or the experiences we have in the laboratory—and our understanding of what the world actually is? That understanding may be in the form of an informal model or, as in physics, a very formal theory.

I'd like to start by speaking about traditional, classical science from 1600 to 1900. You spoke yesterday about how in Buddhist science you have the elements of earth, water, air, fire, and a fifth, space. In ancient Greece and the Middle Ages, the analysis of all objects into five elements is very similar. Later, scientists studied more closely the material objects of the visible world and their behavior. For example, a coin drops or a stone falls. We can describe the fall of such solid objects with great precision, and if we do so, we discover certain regularities or patterns in the way they fall. This was first done in the seventeenth century in Italy by Galileo and became the foundation for the new science of that day. It was a science concerned with exact measurements of objects you could observe in motion, observe with your own eyes. From those observations a pattern was discovered. The pattern could then be described mathematically and became understood as a law of nature; but it was a law that could be verified through visual observation and experimentation.

You could also ask a question on another level: Why does the stone fall? With this, one enters into another kind of analysis. In physics there have traditionally been two levels of analysis. The first level addresses rules and patterns: How, exactly, does the stone fall? The second level addresses causes: Why does the stone fall? What causes these patterns?

Some phenomena are more difficult to observe than others. As science matured during the course of the seventeenth and eighteenth centuries, it began to address more subtle phenomena that could not be seen with the eye or whose causes were not obvious. One set of phenomena that presented a mysterious puzzle for classical science was the production of color. Why is it that colors arise? How do we see colors? Let me demonstrate.

Where Do Colors Come From?

I handed the Dalai Lama a glass prism and showed him how to hold it close to his eyes to produce the right effects. I asked him to look through it to the table below, where I had arranged a set of black and white cards in different configurations. The first arrangement had the white card above the black one (figure 7.1a).

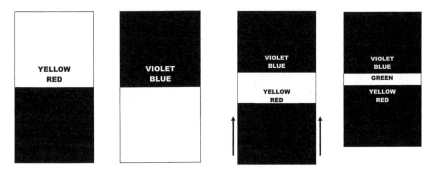

Figure 7.1 (*a, left*) When viewed through a prism, the edges between adjacent black and white areas show warm and cool colors. (*b, right*) If the two edges are brought together and viewed through a prism, green arises.

ARTHUR ZAJONC: What do you see at the place where the black and white meet?

DALAI LAMA: Red.

ARTHUR ZAJONC: Good. [*I now exchanged the places of the cards so that the black was above the white.*] Now what do you see?

DALAI LAMA: Blue. That's weird.

I then added a second black card, arranging the three cards above one another: black, white, black. Finally I moved the bottom black card up slowly to gradually create a narrow slit of white against a black field (figure 7.1b).

ARTHUR ZAJONC: Now watch carefully, in the middle. Do you see a new color?

DALAI LAMA: Yes, half green, half red.

ARTHUR ZAJONC: These same experiments were done by Isaac Newton in his laboratory in Cambridge and by many scientists afterward. It is weird; it's puzzling. Where do these colors come from?

DALAI LAMA: Do I really see the color over there on the paper or in the prism?

ARTHUR ZAJONC: That's exactly the question: Where are the colors? Are the colors preexistent in the light, and do you use the prism to reveal them? An alternative explanation would be that the light is without color and the prism produces the colors. It is difficult to determine which explanation is true. The pattern observed was clear: In one configuration you saw red; in the other case you saw blue; when I brought the two black cards together to make a narrow band of white, you saw green. There are many similar patterns, and there are rules for the production of color by a prism. We don't understand the causes through observation. For this we need another level of analysis. This is where we begin to move away from direct experience.

DALAI LAMA: Does the shape of the prism have any bearing on what color we see?

ARTHUR ZAJONC: It has to be a prism, but the sides can be at different angles.

DALAI LAMA: Does it have to be three-sided?

ARTHUR ZAJONC: No, there are similar effects when any change takes place in the medium through which light travels. Air is one medium, for example, and another medium could be glass or water. The fundamental requirement is a change between the one medium and another.

DALAI LAMA: Is this the same phenomenon when you see a rainbow?

ARTHUR ZAJONC: Yes; there you have small droplets of water falling in the sky. The sunlight behind you enters the droplets and bounces around inside them, and then you see the same effect produced in this more complicated way in the sky.

The point I would like to emphasize is that at one level of analysis one has the patterns and rules. At the next level of analysis, one searches for causes.

What Is Electricity?

For my second illustration of the relationship between observation and theory I selected a simple electrical demonstration. I had a small white ball with a battery, light, and buzzer inside. Two electrodes were on the outside.

ARTHUR ZAJONC: I'd like to give another small demonstration, which is even more puzzling in some ways. This little thing that looks like a Ping-Pong ball has strips of metal on each side. Inside there is a battery, a light bulb, and a little electrical circuit. With a little luck, if I touch the ball on two sides, it lights up.

I held the ball in my hand, touching the electrodes with two fingers, thus completing the circuit. The ball lit up and made a buzzing sound.

ARTHUR ZAJONC: If I take one hand away it doesn't light up. Now I'll ask for David's assistance. He will touch one electrode, and you will touch the other. We can now join hands all the way around the room. You have to make a good strong handshake to complete the circuit.

The participants around the table held hands to form a ring, and the Dalai Lama touched the ball to close the circuit. After a moment, as the handshakes were adjusted, the ball lit up and buzzed, to laughter all around.

ARTHUR ZAJONC: At one level of analysis, the pattern is that when everyone holds hands, and the ring is complete, then the light comes on. When the ring is broken, there is no light. Then there is another level of analysis: Why does this take place? We don't feel anything happening to us; we don't see anything with our eyes, and yet, clearly, there must be a cause. What is the nature of this cause? This is the beginning of a discussion about electricity.

DALAI LAMA: What's the voltage that gets through?

ARTHUR ZAJONC: Very small, about one and a half volts. If it was very high, you would feel it. In the eighteenth century, the French king Louis XIV liked this experiment very much. But he always used an extremely high voltage provided by a Leyden jar, and of course he would not be in the circle himself. He would have his guards hold hands and then instruct them at the last moment to touch the contacts of the battery. They would fly across the room. But I brought a very small battery for Your Holiness.

The mystery still remains: Why is it that we see the colors? Why is it that we see the light and hear the sound? To answer, we begin to move away from direct experience and to imagine or hypothesize the existence of an unseen world. We make an assumption that there are underlying causes and mechanisms that give rise to these two phenomena. Let us focus on the example of the ball. In this case the first hypothesis was that there was a fluid, similar to water but subtler, that would flow around the entire circuit. When this flow was broken, then the ball would not light. When the circuit was complete, then the ball would light. This "fluid" was called electricity, and since then we have discovered more about its nature. We now believe, based on experiments, that there are fundamental particles called electrons, which flow from one side, through all of us making the ring, to complete the circuit on the other side and light the ball. But what is the nature of these electrons? Is it possible to see them? Is there some way that we can gain a direct experience of them? Or must we always infer their existence indirectly?

I would like to show a couple of overhead transparencies that give experimental evidence, not for the electron, but for the atom. We can ask similar questions about the substance of this cloth or anything else. What is its nature? What is it made of? As you know, we think of it as made of atoms. The fundamental act of observation always must have three components: a source of light, the object itself we are looking at, and the eye. But now the object is an extremely small particle, an atom, so small that it can't be seen with the naked eye. Are there techniques that will allow us to see even a single atom? Until very recently this was impossible, but in the last ten or fifteen years, we have gained very interesting evidence of the possibility.

Think of an apparatus like a box, into which I put a single atom. There is no air inside; it is a complete vacuum, containing nothing but this single atom. I then use a laser to illuminate it from outside, and I look with my eye, or with the camera, to see if I can see the single atom. In fact, you can put your naked eye to this instrument and see a tiny, tiny pinpoint of light from the single atom. It's very small, but extremely bright. You cannot tell its shape or its size, but you can take a picture of it.[1]

You could ask, Is there a way to make this image larger? Is there some kind of microscope that could enlarge it? Again, until very recently, this was impossible. It is impossible with a normal microscope, which uses lenses to reflect light into the eye. The light itself is too coarse. You need something much more sensitive, much subtler. Very recently, people have invented a microscope that is more like touch. If I close my eyes and I run my hand across the surface of a piece of paper, I can feel the ridge of a fold, even without looking at it. I could make a map: It's very smooth, and then it rises, and then it falls. In a similar way, scientists at the IBM laboratories have created a very smooth surface on which they can place individual atoms. They can position atoms of xenon in much the same way that you can lift and move a piece of dust by barely touching it. They then take a pinpoint and electrically measure the distance between the pin's tip and the surface. By moving the pin back and forth, the atom appears like a mountain when an image of the measurements is traced on a television screen. You can then see with your eyes what the pin tip feels as it moves. [*See figure 7.2.*]

DALAI LAMA: Aren't the atoms always in a constant dynamic flux, jiggling around?

ARTHUR ZAJONC: Yes, but by making them very cold you can reduce this motion, and then they will stick to the surface of the metal.

DALAI LAMA: These atoms seem rather obedient, perhaps because they are so cold. [*Laughter broke out at this remark.*]

ARTHUR ZAJONC: They are very cold, and because of the vacuum there is nothing around them, so there's nothing to push on them.

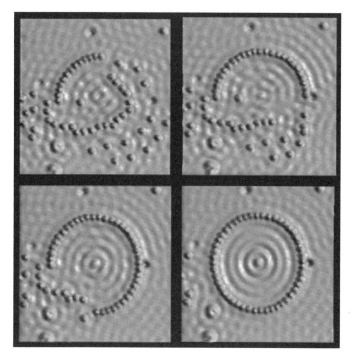

Figure 7.2 Atomic corral. Individual atoms are moved into a circle and imaged by a Scanning Tunneling Microscope. Courtesy of IBM Research, Almaden Research Center. Unauthorized use not permitted.

From Atoms to Attributes

DALAI LAMA: In the Buddhist version of atomic theory, which may be much more gross than the concept of the atom in physics, even the tiniest particle or atom must be composed of eight constituents. These are said to be the four elemental particles—earth, water, fire, and air—and four other derivatives: visual form, tactile form, smell, and taste.

ARTHUR ZAJONC: This is very interesting because it connects to experience. Tactile sense, smell—these are all experiences. In your atomic theory, even the smallest constituents have sense experience components. Your atoms are composite, but composite out of these sense experiences.

DALAI LAMA: I suspect that these elementary particles that form the Buddhist notion of atoms are in fact potentialities. The terms refer to the potentiality or the energy that has the capacity to produce smell, taste, and so forth. Not the smell itself. If each of those elementary particles didn't have that capacity, then the conglomeration of them would also not have it.

This is a most interesting viewpoint. Buddhist atoms are composed not only of the four elements but also of four sensual dimensions. In the West, we tend to see such properties as "emergent." They arise through aggregation. For example, water becomes wet only when many H_2O molecules come together. But other philosophers have protested this idea on exactly the same grounds as Buddhism: You cannot get out what is not already there. The Oxford philosopher Michael Lockwood has argued for years that qualia simply cannot "emerge" from atoms without the atoms themselves having some kind of primitive qualia about them at the outset.

ARTHUR ZAJONC: That's what I wanted to ask. For example, does this white paper have the capacity, the potentiality to evoke white? Or is it itself white?

DALAI LAMA: There is an analogous debate in Buddhist discussions of epistemology. One school, which believes in the objective reality of the external world, holds the concept of *nampha,* or sense data, similar to qualia. There is an unresolved debate as to whether qualia are qualities of the subject or properties of the object or something that emerges as a result of the interaction between the perception and the object. There is probably a correspondence here with the question of color. Physics may offer some clarification here.

ARTHUR ZAJONC: It may help at least to sharpen the questions.

DALAI LAMA: I myself feel that there must be some objective quality of whiteness. Even then, I wouldn't say that the paper itself is white but rather that the color of the paper is white. The whiteness becomes a property of the paper. But then the problematic question arises, What exactly is the paper? Is it the tactile quality?

ALAN WALLACE: There is a peculiar problem for an interpreter here. It is so easy to translate the terms for visual object, sound, smell, or taste. But *tactile* is very awkward. In English we say tactile feeling, or tactile sensation, which places it on the side of the subject, but the Tibetan refers to a tactile object: what is detected by the sense faculty of touch.

GEORGE GREENSTEIN: Is it like an atom of roughness?

ALAN WALLACE: It could be roughness; it could be smoothness.

DALAI LAMA: It is in fact one of the conglomeration of eight elementary particles making up one unit, one atom. One of those eight is the tactile component, the atom's "tactile object." And there are in turn eight types of tactile objects, among which rough and smooth are two.

ARTHUR ZAJONC: What Your Holiness has said connects to my theme of experience versus the theory of modern science. This question arose very early in the development of science in the West: When I see white or red, is it because there is something that could be called whiteness or redness, or is this purely an illusion? Almost all of the scientists and philosophers in the West since Galileo have decided that such things as color are not real, and called them secondary qualities, in contrast with primary qualities, which are the real component parts of the universe. Science has assumed it should focus on the science of primary qualities, which usually includes such attributes as mass, size, and speed of motion. But whiteness, blueness, heaviness—all the qualities, the qualia, the experiences, are only secondary and derivative.

I would like to ask, though I cannot immediately answer, whether it is possible to have a science of qualities, of experiences, as opposed to a science of only quantities of mass, motion, and size. Is it possible to have a science where whiteness and blueness still have a reality and are not only derivative? This, for me, is an important question, though I would like to hold it until the end.

Images of Atoms

The translators had a long discussion in Tibetan with His Holiness to clarify the distinction between primary and secondary qualities. Alan Wallace then presented a question from the Dalai Lama: "When you go down to the minutest level, do you find only round things? Or do you find cubicle things?" Alan explained that he had reminded His Holiness of a comment made earlier in the conference, that at the level of photons there is no spatial dimension and that electrons have mass but no size. I used this as the opportunity to show a set of computer-generated images of atomic structure for different states of hydrogen.

ARTHUR ZAJONC: I'd like to use your question, whether atoms have structure and shape, as a transition to the next step. The following image is not a real photograph but a drawing based on atomic theory. It shows the atom (in this case, a particular state of hydrogen) having a distinctive shape. Although not a direct experimental image, the theory on which it is based has been substantiated by many experiments. There are now experiments in which you can actually visualize these forms through a special imaging apparatus. In the simplest case, you have a sphere. But you can take the same exact atom and prepare it slightly differently, with more energy, so that it takes on other shapes, such as the double-sphere shape. All of the resulting shapes are the shapes of the same atom, but with increasing amounts of energy. [*At this point I showed him a set of figures that diagram the differing shapes of atomic orbitals.*]

DALAI LAMA: It's still one atom; it's not breaking into different parts?

ARTHUR ZAJONC: It's not breaking into parts. This is the mystery.

DALAI LAMA: By increasing the energy, you mean raising the temperature?

ARTHUR ZAJONC: Temperature is a concept that only has meaning for many atoms, not single atoms. You can increase the energy by colliding. Let me describe our picture of the atom to begin with. The simplest atom is thought to have two parts: a positively charged central nucleus and a negatively charged electron. Where is the electron situated? In the simplest case, there is a kind of spherical distribution. But as soon as you increase the energy, through a collision or through incident light, then it moves to another distribution. The mystery is that the single electron, which is one particle, has these shapes that look segmented. Here we enter into a nonclassical concept. How can one thing have a composite shape?

ALAN WALLACE: And you did say before that an electron has no shape.

ARTHUR ZAJONC: An electron has no size and therefore no shape. So, what is it we are looking at when we see this? It's a subtle quantum mechanical question. How do we think about a simple object, which in isolation has no size or shape, although it does have mass and charge? Let me tell you how these pictures are made. One runs many, many experiments, and in each experiment, when it is observed, the electron appears in a different place. When you consider the many different results cumulatively, the electron will have these distributions. For example, in a case where the electron appears in two regions, it appears half the time in one region and half the time in the other region. The patterns are an inherent property of the atom. This is uniquely a quantum phenomenon; these structures of the atom only arise through quantum mechanics.

DALAI LAMA: When you take an atom and bring it down to very low temperature, then it remains more or less still, but is the movement of an electron affected at all by temperature?

ARTHUR ZAJONC: I think that one can excite all of these states even at very low temperatures.

DALAI LAMA: When you described how the atom could be picked up like a piece of lint, this phenomenon is still purely classical physics, right?

ARTHUR ZAJONC: It looks like a classical phenomenon. There is a force involved, and when you move the atom to a new place, it stays there. The objects look relatively normal.

DALAI LAMA: And it is affected by temperature. Whereas, when you are dealing with the movement of electrons, now temperature is irrelevant, and it really is a quantum phenomenon.

ARTHUR ZAJONC: That is basically true. Your Holiness seems quite interested in the question of temperature and motion. I'd like to introduce the concept of zero-point motion. This means that when you bring the temperature down even to absolute zero, the atom or electron will have some residual motion or energy. This will be important in tomorrow's discussion of cosmology when we consider how the big bang could be ignited if, in the initial phase, there is no temperature in the normal sense.

DALAI LAMA: What is the defining characteristic of an atom? We have 110 different types; what is the common denominator?

ARTHUR ZAJONC: That's an interesting question.

DALAI LAMA: Since you pause there, it seems as if the physicists' approach and the Buddhist approach are a little bit different. When Buddhists make such a classification, they will say, first of all define an atom. Having established that, then they will say, now here are the various varieties with that classification.

ARTHUR ZAJONC: Historically, before the concept of atoms existed, there were only four elements (earth, water, air, fire) and the fifth, the quintessence.[2] Then people began to discover that materials had different properties. So, they asked themselves, how many types of fundamental different materials are there, and what are chemical compounds composed of? This was a long investigation. Some chemicals were quite similar. For example, some were gases, transparent and colorless. Others were metals, and there were slightly different types of metals. So, to begin with, scientists of the seventeenth century proceeded by observation and experiment. They still had no real concept of the atom or of elements in our sense. Then, as they began to collect more and more data about more and more elements, and combine them with philosophical ideas concerning the minimal division of matter, the concept of the atom arose. It still seems now, to the best of our knowledge, to be a good concept, although our current idea of the atom is vastly more precise than those first conceptions.

The modern atom has a positively charged nucleus. The nucleus itself is comprised of two types of particles: positively charged protons and neutrons, which carry no charge. Around them is a number of electrons. The number of protons in the center is exactly the same as the number of electrons on the outside. With this knowledge, you can answer the question very clearly. An atom can have one proton and one electron or two protons and two electrons, and so on up to 112. Each one of these is a different element. Each one of these atoms has a different set of properties. The very lightest atom with the least mass has one proton and one electron, hydrogen. The next one, helium, has two protons, two neutrons and two electrons, roughly four times as much mass, and so on. The different properties that result from the different atomic structures correspond to, and help us understand, the properties that were first observed experimentally.

DALAI LAMA: Could you explain why there are only 112? Why can't you have 300 or 3,000 protons in an atom?

ARTHUR ZAJONC: Very heavy atoms are very fragile. They fall apart. When they become so big, the forces that bind them together are too weak to hold all of the aggregate particles together.

DALAI LAMA: Are these classifications of elements universal? Is it exactly the same in other galaxies?

ARTHUR ZAJONC: This is a very important question. One could imagine that it might be different. But when we make observations of the stars with telescopes, we can see exactly the same phenomena that we see in experiments in the laboratory. So we believe, even if we can't travel there, that the same substances exist out in the far galaxies.

DALAI LAMA: Where do quarks fit in here? On my first visit to CERN in Switzerland, I learned about quarks and they fascinate me. [*CERN is an international research center and accelerator complex outside Geneva.*]

ANTON ZEILINGER: Can I add something? Atoms, obviously, have more constituents than just the protons and electrons Arthur has just been telling you about, and he will tell you about quarks immediately. But what distinguishes atoms is that they are the smallest pieces, which cannot be taken apart any further by chemical methods. We can look at smaller pieces, using physical methods, but we cannot take the atom apart with chemical methods.

ARTHUR ZAJONC: You can work with atoms in gentle ways, and they are very stable. But if you apply a strong force, you can pull them apart. If you pull them apart, you discover first that there is a nucleus on the inside, with electrons circulating around it. If you look inside of the nucleus, you find not just protons but also another class of particles called neutrons. Then you can ask whether the protons and neutrons have a composite nature, or are they like the electron, which has a simple, pure nature. An electron seems to have no constituent parts. We now know that this is not the case for the proton and the neutron. If you collide protons, or neutrons, together you find they seem to have an interior structure, which is made up of quarks.

There are three distinct classes of particles. One, exemplified by protons and neutrons, is composite and has quarks on the inside. A second class of particles [*leptons*], of which the electron is an example, seems not to be composite. Photons (and other particles like them) are a third class of particles called bosons.

Unfamiliar Attributes of Fundamental Particles

ARTHUR ZAJONC: I'd like now to return to, and underline, some of the newer and more challenging, fundamental ideas of quantum mechanics that Anton and David introduced. Imagine a situation where a child grows up on a distant island where no bird has ever been seen. A sailor comes and wishes to describe to that person the nature of this creature. If he were to say it flies through the air, this means nothing. He could say perhaps that it is like a stone, which when thrown, moves through the air. But the stone, of course, always falls. So he could say it is like the thrown stone in that it moves through the air, but it is not like the stone in that it does not fall. The child may ask, Does it have arms and legs as I have? Well, it doesn't have arms and legs, but it has wings and feet. What are these wings? Well, they are like arms, but they have feathers. What are these feathers? Well, they are like leaves . . . and so on. I believe personally that we are in a somewhat similar situation in quantum mechanics. We have moved from objects on a familiar scale to objects that are extremely small and subtle. Some of their attributes are much like the attributes of classical objects. They have mass; they can move. Other attributes are quite different, and we don't know how to think of them. Different people use different strategies when they meet such an obstacle. Let me remind you what some of the challenges are.

I'd like to go back to the metaphor of the two glasses. Classically, the particle can be in one glass or the other. One glass contains the particle and the other is empty. This is not a problem for us. But in quantum mechanics, we have a new concept called superposition. I can prepare a system, using, for example, two tiny, tiny boxes and one electron, where the electron is in a superposition state. . . . And like the bird, I don't know how to explain it to you. The electron has, in some sense, a shared existence in these two boxes. It does not have a simple, definite, classical position in one box or the other. This is a wrong way to think.

How do we know that this is true, that it isn't just physicists being careless or ignorant? It is possible to do experiments where ignorance cannot be used as an explanation, where it is only possible to give an account of the experimental results if I allow this state of superposition. If I only allow classical states, one box full and the other empty, then there is no possible way of explaining the results.

DALAI LAMA: Am I right in thinking that the existence of the object in both boxes is not an observed phenomenon but only something that you assume to make sense of the experimental result? That if you assume anything else, it flies in the face of the empirical evidence?

ARTHUR ZAJONC: Yes. The empirical evidence is such that the only possible theory to account for it has to assume the existence of a superposition state.

DALAI LAMA: Does it make any difference how far apart these two boxes are?

ARTHUR ZAJONC: In principle, no. In practicality, it can be much more difficult to prove one has a superposition state if the boxes are far away.

DALAI LAMA: So the electron could manifestly be in two boxes, one on the moon and one on planet earth?

Alan Wallace comments that he and Thupten Jinpa explained that it was not actually manifest, and he asks Arthur for more clarification.

ARTHUR ZAJONC: The simplest evidence is the experiment that Anton did here on this table. He said he could do the experiment with one quantum, one photon, or one electron at a time — with a simple, noncomposite object. We cannot break it in half. You take this object, and you allow it to move through the apparatus. In this case, he used two paths instead of two glasses. When you think normally, you say the photon went this way or it went that way. You can't have this simple object, which is indivisible, break apart and somehow go both ways.

DALAI LAMA: When you set up the experiment and you send the photon through the slit, until it is registered on the detector one could allow the possibility that it went either this way or that way. Once it is registered on the detector, can we trace a path and say it has come through that slit, as opposed to the other one?

ARTHUR ZAJONC: This is a very important question. To obtain the experimental result of the stripes, it's absolutely essential that you cannot tell which path the photon took. If you have a way of measuring the path by observing which of the two slits it came through, then you will not see the striped pattern on the screen. If you set up the experiment in such a way that it is impossible to know, then an interference phenomenon arises, in which single quanta can create the stripes of the interference pattern. This is the measurement paradox: If you determine which path the photon takes, then the interference pattern disappears.

Relation between Scientific Knowledge and Human Experience

There is a discussion in Tibetan between the translators and the Dalai Lama, who is shaking his head in consternation. Alan comments that His Holiness is realizing just how weird it is.

How Do Photons Exist?

ARTHUR ZAJONC: This is very, very interesting, eh?

ANTON ZEILINGER: Because of this very situation, the famous American physicist Richard Feynman, who got the Nobel Prize for inventing one of the versions of quantum physics, said that nobody today understands quantum effects.

DALAI LAMA: In the same experiment, does that photon have a continuous path, or does it not?

ARTHUR ZAJONC: This is a matter of interpretation.

DALAI LAMA: Is there some continuity to its existence? Does it have duration?

ARTHUR ZAJONC: I would say yes.

ANTON ZEILINGER: I would say no. [*This little exchange brought a round of laughter.*]

DALAI LAMA: If you cannot establish a duration for a photon, there's no way you can establish a speed of light, can you?

ARTHUR ZAJONC: This is an important question of interpretation. To me, it makes good sense to speak of the photon as having a continued existence. That way of thinking is consistent with every experiment. If you allow that the photon, or the electron, by nature has that continued existence, then its own intrinsic nature is very strange, and believing this has a big impact on the way you see the world. If you say that it has no continued existence—that only the source, the detector, and certain events exist, and there is nothing that one can say about the particle's intervening existence—that is an easy way to avoid the impact of quantum mechanics. The effects are interesting, but they have no ontological significance. They don't make a statement about reality. For me, I think these experiments make statements about the way the world is.

I'm also a practical person. If this is the way the world is, then maybe I can build a machine that is made of these parts—in the same way that if the world is made of gasoline and steel and so on, I can build an engine. If it's made of electricity, I can build circuits. I don't need to know the ultimate nature of electricity, or anything else, to make it work. But in this practical way, I should be able to build a machine that uses the strange realities of the superposition state of photons, electrons, or other particles. And there is just such a new technology, a computer using quantum states. Your Holiness may know that all modern computers work on a very simple basis, using two states, typically designated by the numbers 0 and 1. It is now possible to build a quantum computer, which not only has these two states but also has the possibility of a third superposition state. Not only is it possible, but also it allows an entirely new set of computational problems to be solved. It's a very, very powerful kind of a computer. We have not yet built the full instrument, but the fundamental parts are being developed, both in theory and experimentally. Anton has been involved in some of these experiments.

So, we enter a new territory with a new set of phenomena. These new phenomena demand new concepts, which are not easily reduced to old concepts. Together these new concepts and the new phenomena can make new practical devices. To me this has a power that affects my life. You asked earlier how the understanding of quantum physics has affected our lives. The normal objects of this world exist in the classical domain of experience, and it is easy to see how they affect us. Their qualities, their qualia, are part of everyday life. There is also another level, of causal mechanisms that are in some ways hidden. We think of these as having no secondary qualities, only primary attributes such as mass and position. Then there is an even more subtle level, where these primary qualities themselves disappear, and new concepts are needed.

Observation as Question

ARTHUR ZAJONC: Let me end this part of the presentation by returning to the issue of the observer's conceptual designation, which we spoke about yesterday. For a physicist and an experimentalist, conceptual designation translates into building an apparatus. I have a thought, a set of conceptual questions. I then make an instrument, a physical apparatus that embodies those questions and will examine the world within that set of concepts. From nature's side, a response is offered to the questions and conceptions embodied in that apparatus. If I ask a different set of questions and create a different apparatus, I get a new set of responses. Through these I build a picture of nature, but always through the conceptual designation that is embodied in the experimental apparatus.

In this sense it's also clear that I cannot remove myself as an observer. In classical physics it is possible to reduce the disturbance caused by observation as far as you would like. Let me give you an analogy. If I wish to see whether my young child is sleeping in bed, then I have to go into the room. If the room is totally dark, I can't see. So I have to turn on the light. If I turn on the light, the child may awake, in which case the child is no longer sleeping. I have disturbed the child by observing the child. But I can go in very quietly and turn on a dim light, as opposed to a bright one. If I reduce the disturbance in this way, at some point I will be moving so carefully that I will not awaken the child. In classical physics, there is no limit to how low the level of my disturbance can be. I can make the light dimmer and dimmer; I can move more and more quietly, without any bottom limit. In quantum mechanics there is a limit, created by the fact that at least one photon must go from the light source to the object and to the observer's eye. There is no way to go below this limit. That is the Heisenberg paradox, that observation will always disturb the system and produce an uncertainty, and there is a lower limit to the size of that disturbance. In these two ways, the observer is very important in quantum mechanics. He is always present in the process.

The observer always disturbs the object observed, but the observer also decides what experiment to run. In these two ways, he or she is always implicated in the experiment. Our intentions and the interests of or our research community are reflected in the experimental design we use. It is a process deeply affected by both social and psychological factors. The very way we see the world is powerfully influenced by these factors. How do we attain objectivity or truth in the face of this fact?

One of the founders of the discipline of neuroscience, Bob Livingston, was in the audience. He had been the scientific coordinator of the second Mind and Life dialogue and edited the conference volume, Consciousness at the Crossroads: Conversations with the Dalai Lama on Brain Science and Buddhism. *He wished to make a contribution from the standpoint of neuroscience, addressing the role of the observer's intentions, past experience, and so forth on observation. I introduced his remarks with two optical illusions.*

ARTHUR ZAJONC: I would like to introduce Bob Livingston's remarks by connecting his ideas to this process of observation. We've talked about the role observation plays in these very subtle realms of modern physics, and we see that it is essential to keep the observer always in mind in the process of quantum mechanics. Now I'd like to return to the other realm of normal human experience and to remark how extremely distant from that realm we are in modern physics. Atoms and electrons are very remote from normal experience, but we treat these things as if they were very real. How real are they? What is the nature of their reality? What

Figure 7.3 (*a, left*) The two small figures are the same physical size on the page. © Exploratorium, www.exploratorium.edu. (*b, right*) Frazer's spiral: The apparent spirals are really circles.

is the role of the observer in this realm? Normally we think of ourselves as being entirely passive. We just open our eyes, and the world appears to us as if we have no part in producing that world. But I have two visual demonstrations, which help us to realize how active we are. The first image [*see figure 7.3a*] shows how we judge relative size. Even though it does not seem as if it could be true, the small figure in the foreground is exactly the same size as the figure in the background. The second illusion is even more compelling. The picture seems to be a set of concentric spirals. But actually if you look at it carefully and trace one of the "spirals," you can convince yourself that it is really a set of concentric circles. [*See figure 7.3b*.]

DALAI LAMA: In Buddhist epistemological discourse, there's an extensive discussion about optical illusions and whether the source of illusion lies in the object or in the visual perception or has something to do with the environment.

ARTHUR ZAJONC: These effects open exactly those questions for Western science, also. What is the role of the person who perceives? Within this context, I would like to give Bob Livingston a little time to speak about the relationship between that which comes to us in observation and that which we bring to the observed.

The Plasticity of Perception

BOB LIVINGSTON: I want to talk very briefly about the brain-mind. In this day of organ transplants, it's possible to replace a damaged heart by taking an intact heart from a man who has been otherwise very badly damaged. It's possible to transplant a lung, a liver, a kidney, even skin or bones. But if you transplant a brain from one individual to another, you're actually making a body transplant because you take the personality and the interpretations of the world from this brain and put it in a new body. Evolution has provided for our being able to see patterns in photons, to see patterns in structure and in tactile objects. We have a sensory apparatus that tells us about the outside world and we make constructs of that world, which enable us to walk up and down stairs and do things in the world with a certain amount of confidence in our perceptual experience.

We think ordinarily of our having sensory nerves, which carry information about visual objects, tastes, olfaction, or other experience into our brains, and we make models of the world in which we survive, design, expand, and elaborate. What we often don't recognize is that all the nerves that come in from sense organs are matched by nerves that go out to the sense organs. The number of nerves that go out to the sense organs varies from 10 to 50 percent of the bundle. There is a conspicuous contribution from the central nervous system to the sense organ.

The nerves coming out from the central nervous system to the retina can affect the impact of light on the sense receptors and can particularly affect the relay within the retina of events excited by the photon bombardment. They can also affect the message that goes back to the central nervous system. Similarly, in the central nervous system, each of the relay projections can be modified from central projections outward. These outward-projecting impulses act in accordance with our past experiences, our expectations, and our purposes. Our past experiences dictate a great deal about what we perceive from our retina, from our auditory apparatus, and so on and make an idiosyncratic experience for us, unique to the individual.

Evolution has given us access to the world and has also given us the power to modify that world experience in accordance with our past experiences and our expectations and our purposes. If we change or have different past experiences, we see things, feel things, experience things differently. If we change our purposes, we can radically change the input in our perception. If we change our expectation, as in an athletic or dance or musical experience, however complicated it may be, we modify the sensory experience. This means that we live in a world in which evolution has contributed and our personal experiences have contributed. We are obliged to accommodate ourselves to a society, and as we grow up that society exercises a lot of discipline over us, making our images conform to the societal imagery. Errors in our individual experience and in our societal experience can be very dangerous if they are in conflict with one another. So, I appreciate what Adam Engel and His Holiness have done to create this dialogue for very important communication between different levels of experience and perceptual understanding of what we are and what we have as potentiality.

DALAI LAMA: What happens to objectivity? If we seriously believe that our very perceptions are heavily structured by past experience, expectation, and purposes, then is there any residual notion of valid objectivity—even intersubjective or interpersonal objectivity? Or is it all gone?

BOB LIVINGSTON: I think that we are searching, individually and collectively, for some stability of imagery, but we can never be positive about it. We search and we can find some ways of anchoring, whether it's in physics or physiology or religious conviction, but we cannot ever be absolutely certain. This is because evolution has given us the grief of freedom, as well as the extraordinary capacity. Our individual experiences are so different from one another that the world consists of a couple of billion people and a couple of billion worlds.

We seem caught in a strange irony. The discoveries of science, which we take to be so certain, seem to undermine that very security. What does happen to objectivity, to true insight, to enlightenment in a world that "consists of a couple of billion people and a couple of billion worlds"? Is the radical subjectivism that Bob Livingston identifies the only way to reconcile the facts of science with our everyday experience? Buddhism has tended to reject radical relativism as leading to nihilism, with all its dire moral implications. What can it put in its place, without a return to fundamentalism or a simplistic realism?

8

Investigating the World, Pondering the Mind

Western science has focused almost exclusively on the external world, developing ever more refined methods of investigating it experimentally and of understanding it through increasingly abstract and comprehensive theories. In the wake of scientific advances have come myriad outer improvements in human life, although often at serious environmental cost. The sciences of the mind—psychology, cognitive science, neuroscience—are relatively young, and often they neglect the subjective experience of the individual in favor of third-person research methods and theoretical accounts similar to those of the physical sciences. By contrast, Buddhist monk-scholars have searched for the inner reasons of human suffering through close attention to human experience itself. They have long recognized the dangers of delusion and distraction, and they have developed an exacting contemplative discipline that allows them reasonable certainty about the results of research into this interior territory. Their motivation has been to alleviate human suffering not so much by technical innovation but by psychological and spiritual practices that work directly on the mind. Why the dramatic difference in emphasis between East and West? Each of the conference participants wished to address this question, which was posed by the Dalai Lama. Their comments became the basis of a fascinating and wide-ranging conversation, contrasting the place of rationality in East and West.

DALAI LAMA: Do you have any idea about why science throughout its history has placed so much emphasis on understanding the nature of the objective world, whereas there does not seem to be equal emphasis on seeking the nature of the observer or the subject. Why is there a fixation on the external?

DAVID FINKELSTEIN: I think of the development of physics as the growth of relativism. Every increase in relativism is an increase in our understanding of the complexity of the observer, of a number of possible observers, and of the influence of the observer upon the phenomenon. Quantum theory is only the latest stage in this growth of relativism. It is by no means the end of our understanding of the importance of the observer or the experimenter. In fact, I think it's no longer proper to speak of the observer because that still carries the old implication of someone who observes from outside the phenomenon without changing it. In quantum theory we only have experimenters, people who do things. One of the lessons of relativism is that one should be very careful about assuming the existence of things that in principle cannot be seen. Evidently one of these things is the universe. It's a physical impossibility, as a matter of principle, for any experimenter to have complete information about the universe. Every experimenter begins by ignoring large parts of the universe, namely, himself or herself. You turn your attention to something by omitting from your attention much more. It is quite possible that the universe will go the way of absolute time or absolute state. I don't think we have reached the end of this growth in relativism.

DALAI LAMA: Given that, at least in a manner of speaking, we have the observed and the observer, why has the overwhelming attention gone to the observed, to the overwhelming exclusion of the observer. Can you explain that?

ARTHUR ZAJONC: This is not a definitive answer, but allow me to give one view of why Western science has historically focused on the external world. Through the Middle Ages, there was not, in fact, such a great focus on the external. Of course, craftsmen needed to know the material world to make buildings, garments, and so on. But they were not scholars or monks. They were people who could not read or write. Around 1600 this changed in Europe, and for the first time scholars began to make things themselves and to take a very deep interest in the material world. For example, Isaac Newton, who was a great scholar and mathematician, made his own telescope. It was a beautiful telescope, better than any craftsman in London could make. Galileo, likewise, made a different type of telescope. So the tradition of the scholar who never made anything, because such craft was impure, more proper for the lower castes, began to change. The scholar became interested in the substance of the world and in bringing his reason into substance. In that endeavor lay great success and great power. As a consequence of those successes, science became an increasing force in Western civilization. It also became a threat to religion in certain circumstances. And so, for example, Galileo was arrested and imprisoned. A tension grew between the inner world of the spirit and religion and the outer world of the mastery of nature. A clear division took shape. The religious world conceded to science: As long as you say nothing about morality and the inner world, then we will give you the outer world. You take mastery of the outer world, and we will take mastery of the inner world. In the beginning, science was very weak and religion was very powerful. But now things have shifted, and the power of science and technology is very great in the West. It has become difficult now to have a conversation between a spiritual leader, such as yourself, and scientific leaders, such as the participants here, who now wish once again to bring these two worlds a little closer.

The Theological Premises behind Our Scientific Attitude

In the audience was a distinguished Harvard historian of science, Anne Harrington. She had been part of a previous Mind and Life meeting in 1995. I invited her to comment on the question posed by the Dalai Lama.

ANNE HARRINGTON: Western science comes out of the Christian tradition, and we have to contend with that. There is a theological premise underlying science: the fundamental premise that creation is comprehensible to the human mind, that nature is rational. The Christian view was that man was created in God's image. That was understood at the time of the scientific revolution to mean that man's rational mind was created in the image of God's rational mind. Having created nature, God had given human beings a rational mind, like His own, that could comprehend nature. In the Christian view, God stands outside of creation. Similarly, to understand creation, the scientific point of view must in some sense be aligned with God's perspective on creation. To comprehend and imitate God's creation, human beings had to take themselves outside of that creation and act as if they were God looking down at it. So we say that the scientific perspective is a "view from nowhere," a God's-eye view from nowhere and from everywhere.

TU WEIMING: There's another dynamic going on that supplements Anne's point. Because the theological position is that God created the world from outside, no human mind, no matter how rational or comprehensive, would ever be able to fully understand God's intention. There is a leap of faith in the Christian community: Since God created the world and man with rationality, human beings would be able to understand the world through rational means, but there is no need for human beings to try to understand the totally unknowable God. This led in the eighteenth century to a very powerful anthropocentrism, which is still very much with us. The European enlightenment movement was an attack on the clerical tradition of religion and led to a separation of religion and reason. On the one side, religion becomes faith that cannot be defended by rationality. On the other side, human beings are seen as able to understand nature through the new experiments and instruments that human beings created. Religion is rejected by science as God-centered, and the notion that we are simply an integral part of nature is also rejected. This was Francis Bacon's position, and it was very powerful. Nature is not going to reveal her secrets to us voluntarily, and so we have to use instruments and interventions to force nature to tell us what she really is. That notion of anthropocentric intervention, combined with the anthropocentric rejection of religion, is what people mean when they talk about secularization.

ARTHUR ZAJONC: This is an important point: In this view God is the infinite, and since we are finite, we can never know the infinite. There is an abyss between us and God. We can, however, know the finite world with the finite mind. This break comes right around the time of the scientific revolution, the division between faith within the religious and spiritual tradition of the West and knowledge concerning the natural world.

DALAI LAMA: Probably in the early stages of this history, all of the scientists were operating within a cultural context where their conditioning was very strongly Christian, but as time went on later generations of scientists would probably consciously disavow any form of theological conditioning in their upbringing.

ARTHUR ZAJONC: This is interesting. Look, for example, at Max Planck, who invented the idea of the photon around 1900. He, like many other scientists of the early twentieth century, subscribed to a religious viewpoint known as neo-orthodoxy, which conceived of a particular arrangement between two worlds. One is the world of nature, and is given to science, and the other is a world of moral life and "ultimate concerns," as theologian Paul Tillich called them, and it is given to religion. Many scientists had a religious life, but it was disconnected from their scientific and rational life. This division was shown by a dramatic example.

After the bombing of Hiroshima and Nagasaki, Karl Barth, Europe's most famous Protestant theologian, was asked to speak with physicists concerning the moral implications of atomic weapons. He refused, as did all of his students. When asked why, Barth said that scientists had one world, he had another, and they had no common ground to speak about. He believed it was a logical impossibility that they, the scientists, could have anything to say about the morality or the immorality of the bomb, even though they had built this device.

Many scientists still subscribe to some version of this view, perhaps not one as radical as Karl Barth's, but they hold a two-realm theory of truth: a moral realm and a scientific realm. More recently, over the last thirty years, we are beginning to see changes. Some scientists are more inclined to disavow all religion and spirituality, viewing the world as purely material. And a few scientists are trying to find ways of bringing both spiritual and physical understandings together. These scientists are a minority, however, within the physics community. Also, of course, religion has become more complicated. Christianity no longer dominates religious life as it did before. Buddhism in particular has had a very powerful effect recently in the United States and in Europe.

GEORGE GREENSTEIN: I'd like to add a different kind of answer to the question. Science asks very limited questions. It asks, "How much does this weigh?" When you answer that question, it doesn't matter who you are. If you are happy or sad, male or female, you still get the same answer. The observer is irrelevant to the question. Anton has shown us an experiment where the observer is very relevant, but for most things in nature, it does not matter who the observer is. It has only been recently, in this century, that we have discovered things for which the observer is relevant and that are now pressing us into these considerations. But until this happened, there was no reason for scientists to care about their spiritual nature or about themselves as observers.

These diverse contributions succeeded, at least in part, in answering the question of why science has focused so much on the external world. In the West, the inner world was given over to religion, which historically has had an adversarial relationship to science. Buddhism, by contrast, had from its beginnings a clear and positive orientation toward knowledge, believing that delusion is a great source of suffering and that insight into the nature of the world and self can break the cycle of suffering. I was reminded of the opening words of the conference, when the Dalai Lama said that one should practice initial skepticism and always remain open. Clearly the two-realm map of reason versus faith—or as Stephen J. Gould has termed it, the doctrine of "non-overlapping magisteria" that divides science from spirit—has never been adopted in Buddhism. For many years I have felt that the neo-orthodox map we have drawn has been artificial and damaging. I was encouraged to find it absent in traditional Buddhist philosophy.

Reshaping the Mind

The conversation took a very different turn after this. Two closely related themes were interwoven. One concerned the possibility of training the mind to broaden and enrich that which humans can experience, and the second concerned the ultimate emptiness of inherent existence. How far can experience take us? And when we try to reach beyond experience to the thing-in-itself, what do we find?

ANTON ZEILINGER: This morning we stumbled upon one of the deepest questions in quantum mechanics, signified by what we call the superposition principle. The question is not really about logical deduction—whether the electron is here, there, both places, or neither—but about how to understand this phenomenon. There was a famous Austrian physicist named Wolfgang Pauli who was known for his sarcastic remarks. When the American mathematician John von Neumann, who was very proud of having calculated some proof, told Pauli that he could actually prove this point, Pauli replied that if physics required nothing more than being able to prove things, then von Neumann would be a great physicist. So, we really have to grapple with concepts now rather than proofs.

ARTHUR ZAJONC: As Anton said, the results of our experiments in quantum physics have proved the superposition principle. But, speaking for myself and for many physicists, we do not understand the superposition principle. I'd like to connect this problem to the problem of experience. Very often our understanding arises through experience. We may have heard the description of a bird, but one day we finally see a bird and then we know with certainty what the meaning of *bird* is. In quantum mechanics, we now have a description of superposition. There is good evidence, but we don't understand it experientially.

If we try to understand this problem from the standpoint of normal sense experience, with the analogy of an object in two glasses, we can't succeed. But is that the only kind of experience we can bring to bear? In Tibetan medicine, there is a technique of reading the pulse. Young students, beginning to study Tibetan medicine, feel almost nothing, or feel something uncertain, ambiguous, and unclear. But through long training, they begin to feel something—maybe just a little at first, and then it goes away, but eventually they train their awareness. For an accomplished master, that same pulse becomes a window, a very accurate, refined way of seeing. Experience can be changed by changing one's mind. Experience always arises in the context of the mind. If the mind is dull, then the experience will be dull. If the mind is very sharp, then the experience will be very clear.

Perhaps we have a dull mind with regard to superposition. Is it possible to school the mind by becoming attentive to these phenomena, just as one could become attentive to the pulse, so that we begin to experience these nonclassical, quantum mechanical states? I think physicists would have different responses to this question. The response of Niels Bohr, and I believe of my dear colleague Anton Zeilinger, would be no, this is not possible: Our awareness is shaped by the physical phenomena of the sense world, and this will limit how we can understand the concepts we bring to even the quantum mechanical phenomena. We will always have to resort to this domain of experience and to the concepts that are derived from it. I, on the other hand, am more optimistic, or at least open to the idea that consciousness is malleable, that the mind can be trained.

ANTON ZEILINGER: The problem is related to the question Your Holiness raised yesterday: How can we validate cognition without any truly existent object? I don't accept Arthur's position, although I admit it is unquestionably possible. I don't accept it because, if I apply quantum analysis and the superposition principle to everything, then everything dissolves and loses its well-defined properties. I must be careful about how I speak and what questions I ask. So to make sense of what I am doing, I choose not to apply my quantum analysis to the pieces of the apparatus. I simply posit them as existing, with well-defined properties. I restrict myself, in the Copenhagen interpretation, to making statements about properties of the quantum objects independent of the apparatus. I have accepted the apparatus beforehand as existing and as well defined.

In the Copenhagen interpretation, due to Bohr, a clear distinction is made between the system under study, with its quantum nature, and the apparatus used in the analysis, which is treated classically.

DALAI LAMA: This seems to be analogous to the distinction between ultimate and conventional truth within the Madyamika view. In terms of ultimate reality, you cannot posit anything at all as existing from its own side, by its own inherent nature—not even that emptiness of inherent existence itself exists. In terms of conventional reality, you can posit the existence of all kinds of attributes, phenomena, interdependent relationships, and so forth. In the conventional mode of engaging with reality, you are satisfied with the nominal status of mere appearances, the way we speak about things normally. Within that realm of appearances, you can make all types of analyses about a phenomenon: its causality, its attributes, the states that it moves through, properties, and so forth. And within that context, it is still possible—in fact it's imperative—to make the distinction between a valid cognition and an invalid cognition. It is possible to make genuine discoveries, and genuine mistakes, within that realm of appearances and conventional phenomena.

But you may not be content with the mere appearances, with the conventional status of a phenomenon. What about its actual nature? What is it *really?* When you start probing beyond the appearances, trying to understand the real nature of the existence of the imputed or the designated entity, this is called an ultimate analysis, seeking the nature of ultimate reality. When you start seeking that, you don't find anything at all. In fact, you find that there is nothing to be found. The very "not finding" of a phenomenon, when you seek it through ultimate analysis, is what is meant by emptiness.

Those two modes of analysis, ultimate and conventional, are mutually incompatible. If you are doing one, you are not doing the other. Nevertheless, they are both made with respect to the same basis. They are both talking about the same phenomena. This seems to be analogous with your mode of discourse here. Insofar as you can penetrate to the very subtle nature of minute particles, it seems there is nothing to find. But if you back up and are content with the macroscopic, the gross phenomena of normal appearances, then you can say a lot. Nevertheless, these two analyses are made with respect to the same thing. The quantum phenomena are here in the same objects, such as this bottle, to which the rules of classical physics apply. You don't have to go someplace else.

Moreover, it seems to be the case that the gross phenomena arise from the subtle. The appearance of the bottle itself stems from that quantum realm. We are not speaking of sequential causality here—first the quantum reality, leading then to the gross reality—but rather the quantum realm is the basis for that which is manifesting in the gross macroscopic world. In Buddhism, emptiness of inherent existence is the basis for the appearances of conventional phenomena. It's not producing it, but it's the basis for it.

PIET HUT: Likewise in physics, not only do we search and not find, but we find that there is no possibility of finding.

DALAI LAMA: It's the same in Buddhism. If the ultimate nature of a phenomenon were to exist, it could be found. You don't find it; therefore it doesn't exist. We have discovered that it is not to be found.

PIET HUT: In physics, the method would be to do an experiment. Then, if there is no theory that can be derived from the experiment that fits the existence of the phenomenon, the theoretical interpretation of the experiment shows you that the phenomenon has no existence. In Madhyamika, what are the roles of experience and theory in reaching the conclusion of nonexistence?

DALAI LAMA: There is an emphasis in Madhyamika on rational analysis, but it's not merely a matter of logic; there's also a very crucial empirical, or experiential, component to it. For example, you may do the analysis with regard to yourself as the object: Do you as an individual exist inherently by your own nature? Are you a truly existent phenomenon? Before you bring logic to bear, you first need to experientially identify what this self is. Do you indeed grasp onto yourself as being truly existent? Do you participate in such a reification of yourself? If so, what is the object of that reification? What is the nature of this truly existent self that you conceive of and that you regard as "you"? You need to seek out experientially, and very vividly, what the nature is of the phenomenon that eventually you will refute. You hold it, almost like putting it into a particle chamber. You isolate it vividly in your mind, and then to that experiential basis you bring the rational analysis. The strategy involves several different logical approaches. For example, you seek out the actual nature of a phenomenon by understanding its dependent origination—the manner in which a phenomenon exists as dependently related events. Another common strategy is to analyze the phenomenon in terms of the whole and its parts—the entity itself and the attributes and components of which it is composed. How does this self exist in relationship to your body and your mind? Is it the same as these components? Is it separate from these components? As you apply this rational analysis, you see that it's not this, not that, and there's no third alternative. If a truly existent self must exist either in this way or that way, and it doesn't exist in either way, then you haven't merely not found it. Rather, you have found that it doesn't exist.

You analyze this and you come to a conviction, a belief, sense of a confidence, but you do not yet ascertain it as an actual inferential cognition. How does your awareness of that reality shift? I can believe that George is from the United States without knowing it, and I may happen to be right. But that's not the same as knowing it. So you keep probing your conviction experientially, as well as rationally, familiarizing yourself with it and going deeper and deeper into it. Eventually you move by means of inference beyond this conceptual realization to a purely perceptual or purely experiential realization. And that is really the goal. So, the strategy is to move from belief, to inferential valid cognition, to perceptual or purely experiential valid cognition.

Valid Cognition: From Inference to Experience

The sequence outlined above seems central to the Buddhist understanding of valid cognition: from belief to valid inference and finally to valid experiential cognition. I hoped for more detail, so I asked the Dalai Lama to go further.

ARTHUR ZAJONC: You contrasted two very different modes of analysis, one conventional analysis, which allows us to speak about the normal objects of this world and to have a normal existence, the other a very deep, probing analysis that ultimately shows the emptiness of things in and of themselves. But this seems like a very coarse gradation: It's one or the other. In the sense world, there are many objects I can see with my senses, but perhaps there are also conventional realities that are more subtle and that elude detection by the senses.

DALAI LAMA: Perhaps it may be useful to look at another Buddhist classification of the three realms of knowledge. One is the domain of manifest, or evident, phenomena, which includes all objects that can be empirically experienced and perceptually known. The second realm is described as obscured or hidden phenomena, which you cannot directly experience or know perceptually but which you infer on the basis of empirical evidence. There is a third realm of knowledge known as extremely hidden phenomena. They are said to be beyond the capacity of both inference and direct experience for an ordinary person. Perhaps the only avenue to this knowledge that we have is a third person's testimony. This classification is presented from the perspective of an ordinary, untrained human being. With training, one may be able to perceptually engage with realms beyond manifest phenomena, but not as a beginner.

GEORGE GREENSTEIN: Just to make sure I understand. . . . The first level of knowledge is what I see. The second level would be looking through a microscope. Is that right?

DALAI LAMA: No, not quite. An example of a hidden phenomenon is the momentarily changing nature of this bottle.

ALAN WALLACE: Brownian motion is another example of a hidden phenomenon. We didn't at first see the atoms roving around in their chaotic fashion. It was something that could only be known by inference. Then, maybe seventy years later, a microscope was actually developed that could let you see the atoms moving around. But at the beginning it was a hidden phenomenon, only accessible through inference.

GEORGE GREENSTEIN: How would you classify what I see when I look through a microscope?

DALAI LAMA: This would still be an evident phenomenon. I'm trying to recall the classification of five different types of form. There is one type of form that you can see with the naked eye and other types of form that can be perceived, but not in the ordinary manner with the visual sense. For example, the images that you see in a dream are certainly not seen with the eyes. There are other classifications of form that are not seen with the eyes, but I can't recall them now.

His Holiness consulted with the monks in the audience, to no avail. "They don't remember either," he said, which brought a laugh to all of us.

ARTHUR ZAJONC: Well, I couldn't remember the details of the quark model either, so you're okay.

DALAI LAMA: The three classifications of knowledge—evident, hidden, and very hidden phenomena—can all be found within the realm of conventional reality. Moreover, all of these pertain to a subject. And all of them depend on context. None of them are intrinsically evident, hidden, or extremely hidden. What about ultimate reality, or emptiness? That, in fact, is in the second category. Ultimate reality is hidden: not evident, but not extremely hidden. It's sometimes called slightly hidden. I want to emphasize that conventional reality does not correspond only to the realm of manifest.

DAVID FINKELSTEIN: I didn't quite understand the third realm, the very hidden. Could you give an example or a fuller definition? You mentioned the possibility of learning about it from a third person. Is it manifest to the third person?

DALAI LAMA: Yes. An example cited very commonly in Buddhist teachings is the very subtle working of karma, the ethical significance of an action, whether good, bad, or neutral. From a Buddhist perspective, we would say that only a Buddha has a direct perception of the extremely subtle consequences that result from an action, perhaps over hundreds of lifetimes. For everybody else, all other sentient beings, this is extremely hidden. It sounds rather sectarian to say "only a Buddha," but to state this more generally, the extremely subtle relationships of actions and their long-term consequences remain extremely concealed until one has removed even the most subtle obscurations of the mind. For someone who has removed even the most subtle cognitive obscurations, those relationships are evident.

For a more accessible example, imagine that I am speaking on the telephone with a friend, and right outside my window I see a little child falling off a bicycle and crying. The child is in pain. For the child, the pain is an evident phenomenon. For me, the pain is inferred: I only see the outer display. I don't see the child's pain because I'm not clairvoyant, but I infer the child's pain because the child is crying. And then I tell my friend on the phone about what I have witnessed. For that person, the child's pain is not evident and cannot be inferred logically. But he will take me as an authority, knowing that the child is in pain, based on my testimony. If the person simply believes me without knowing anything about me, that's mere belief; it's not knowledge. Maybe I'm a pathological liar or deluded or hallucinating. There are criteria for this: If the person has investigated my status and has come to the conclusion that I am an authority on what I'm talking about, then that person can claim knowledge based on having heard from an authority. That's not so easy. It's a lot more difficult than simply having a belief. Another example is my own birthday. I know when my birthday is only because my mother tells me, but I don't know from direct experience or from inference. My mother told me I was born on this date, so I believe her because she is an authority on this.

GEORGE GREENSTEIN: On Monday, Anton showed us the apparatus where we saw an interference pattern. That's the first level of knowledge. From that we infer superposition. That's the second level of knowledge. What would the third level of knowledge be in this case?

DALAI LAMA: Probably the third level of knowledge would be the reasons for the fact that we were doing this experiment, on that particular day, at that particular instant, in this particular gathering. Why did that happen, in all of its specificity? Was it a pure coincidence, or was it due to our previous individual karma, our common actions, or some other factors?

ARTHUR ZAJONC: You said that it is possible to move from inferential knowledge to a manifest, experientially based knowledge. In the example of the child crying, I can infer the child is in pain because I see the symptoms. But how is it possible for me to experience her pain, to know it experientially? It requires changing my consciousness. As long as my consciousness is based only on the senses, this is impossible. You mentioned clairvoyance—are you proposing a way of knowing experientially that is other than the consciousness of sense reality?

DALAI LAMA: If you can only infer the child's pain, your inference doesn't turn into a direct perception just by focusing on that for a long time. However, in the Buddhist view, there are other methodologies, such as developing very profound *samadhi*, that make it possible to develop a clairvoyance that would give you direct access to the child's pain.

Perhaps here we need to take into account the complexity of the types of knowledge. For example, in the case of emptiness that we spoke about earlier, it is conceivable from the Buddhist point of view that inferential knowledge can culminate in some kind of experiential knowledge through a process of familiarization. A Buddhist meditator's understanding of the nature of existence as suffering initially may be very intellectual and inferential, but as a result of prolonged meditation and practice, the intellectual knowledge can culminate in an experiential knowledge of that suffering. But this does not mean that every kind of inferential knowledge can culminate in experiential knowledge.

In Buddhist discourse, we distinguish between two types of cognitive activity. One type involves a kind of knowledge, such as the knowledge of emptiness or the knowledge of the suffering nature of existence. Knowledge of this type begins inferentially and then culminates in experience. But there are other types of cognitive activity involved in the Buddhist path, which are generally labeled as skillful means. These include the more altruistic dimension of the spiritual path, such as meditation for developing the compassionate potential of heart. There is no suggestion that this activity is inferential cognition that at some point turns into direct experience. It is not even a mode of knowledge. However, in the initial stage it could be described as a simulated state, which as a result of prolonged practice becomes spontaneous. It is effortful at first—you meditate in a certain way, and a sense of compassion arises, but when you stop thinking that way then it goes away again. But through sustained meditation and familiarization with that state, then eventually it arises effortlessly. At that point it is called uncontrived.

ARTHUR ZAJONC: Your Holiness, I'd like to go back to my original point regarding the relationship of experience to the knowledge of quantum physics. We infer quantum superposition; it's not something that is a matter of direct experience. But is it possible that some methodology, whatever that might be, could culminate in a direct experience of superposition? It is not clear to me whether all inferred phenomena have the possibility of becoming experiential knowledge or whether there is a class that always remain inferred.

DALAI LAMA: In general, from the Buddhist point of view, all forms of inferential knowledge have the potential to culminate in direct experience. Anything that can be known by inference can, sooner or later, be known perceptually or experientially.

Of course, I was pleased by this answer because it corresponded with my own tentative opinion on the matter, Anton's qualms not withstanding. I said as much at the moment, which brought another round of laughter. It had become clear that all properties, manifest or hidden, even quantum superposition, are part of what Buddhism calls conventional reality. Scientific analysis largely, or perhaps exclusively, deals with this level. When one shifts to the complementary mode of analysis, asking after the ultimate intrinsic existence, one finds only emptiness. On the one hand, this makes sense to me. I had rejected metaphysical realism, which asserts the contrary; but like others, I am repelled by the apparent ethical consequences of this position. I was anxious to probe deeper.

Emptiness and Ultimate Reality

ARTHUR ZAJONC: Your Holiness, in terms of emptiness, how do you avoid complete relativism? When you make the deep analysis, and you find nothing, how do you avoid the problem of assuming reality is simply a mental construction?

DALAI LAMA: This relates to a point that was raised yesterday, that within the Madhyamika context, where there is no absolute reference, there can be no strict correspondence theory. Nevertheless, within the context of conventional reality, we do make distinctions between valid cognition and invalid cognition. In other words, it's not arbitrary or whimsical. And reality is not determined by consensus because a great many people can hold a false view.

So, what do you do? To take an example, imagine you are walking through a forest in the twilight, and you see what is in fact is a coiled rope on the ground. You don't see it clearly, and you think it's a snake. You don't know whether that's a valid cognition or not. It looks like it, but you don't know for sure. You investigate more carefully—you are not worried about its ultimate mode of existence at this point; you are just investigating its conventional nature, exactly like in scientific research. If, as you probe more deeply, your initial conclusion is confirmed, then it stands. But if, upon very careful further analysis, you find out that your initial conclusion was false because it is invalidated by a more precise conclusion, then within that context the latter cognition is valid in reference to the former. But now you have this latter cognition, which says something else. This, too, can be subjected to even more critical analysis, still within conventional reality. It is very, very similar to scientific research.

ARTHUR ZAJONC: The question still remains. You said earlier, for example, that both the conventional and the deeper, ultimate analysis of this bottle are made on the basis of the same thing. Did you mean that the same phenomenon is the basis or that something comes from the bottle's own side, so to speak, from nature itself? Is there really no objective grounding for phenomena?

DALAI LAMA: Not ultimately, from its own purely objective side. But even if you take the bottle conventionally as it appears, without subjecting it to an ultimate analysis, perhaps we should not see the conventional realm as unidimensional. Perhaps even within the conventional realm there are different levels. At the very gross level, for example, if you ask another person to buy a bottle of water, neither of you would consider the particular molecular structure of the bottle. Both you and the person you have asked to buy the bottle know what you mean on the level of transactional usage of language. If you go deeper and ask what exactly the bottle is made of, in some sense we are now looking at a much subtler level. But the physical constituents of the bottle at the atomic level are still, from the Buddhist point of view, within the realm of conventional reality. However, you said that in quantum analysis you reach a point where the very existence of bottle begins to disappear. This seems in some way parallel to the Buddhist analysis of emptiness, where the end result arrives at a pure negation. You do not posit or affirm anything in the aftermath of that negation. This is important from the Madhyamika point of view; it is called nonimplicative, or simple, negation. The process may not be *via negativa*, but what you arrive at is a negation.

Anton, you spoke about not finding anything as a result of pursuing the quantum analysis. Would you say that that is a form of negation?

ANTON ZEILINGER: Yes. It is a negation of the existence of something that I posed as a question.

DALAI LAMA: It's ironic that analysis approached purely from a physicist's point of view, and confined to physical phenomena, seems to reach a point where it may just be opening the door to Buddhist emptiness. The irony is that there seems to be a total negligence or nonacceptance of negative phenomena in the process of your analysis, a view of physical reality that seems to address only the affirmative, only what it is. And yet through that process you end up at a point of negation.

PIET HUT: The absence of hidden variables is a form of negation.

It is interesting that in physics there are two ways to lead to a valid conclusion. Einstein's theory of relativity is an example of one. First, he did a thought experiment in his imagination, which led him to a new conclusion. Then it was verified experimentally in the laboratory. But quantum theory was unexpected. First, very puzzling experimental results appeared, and then the theory was found later. Some things are found through a clever idea or the imagination of a genius, and some are presented unexpectedly by nature. Sometimes experiment leads, and sometimes theory leads, but in both cases, you make sure you are correct by using the other approach to verify your conclusion. I'm very curious whether in Madhyamika and in other forms of Buddhist meditation there are also two distinct modes and what the relation is between them.

DALAI LAMA: There are, in fact, two different approaches that are comparable to what you have suggested. In one, you establish a structure of theoretical insight and then use that as the platform for entering into the meditative experience. The experience emerges from the prior view. Another strategy starts with the experience. You enter into the meditative experience, and out of your experience you articulate a view.

Generally speaking, insofar as you are seeking the view of emptiness, the ultimate nature of reality, both to the exoteric Sutrayana and the more esoteric Vajrayana practice, both emphasize that there is no other more effective strategy than first establishing the view by means of reasoning and theoretical analysis and using that as the platform to go beyond into the actual experience. However, in Dzog Chen, or the Great Perfection, and other modes unique to Vajrayana, there are instances in which, by the sheer power of total reverence and a very, very high degree of spiritual maturation, you are poised or ready to gain genuine experience without first having a theoretical framework. Then, following that genuine experience, you may seek to articulate what you've already experienced in the form of a theory.

The answer is yes, there are both. But in the second case, Buddhists would argue that the person's spiritual maturity was a consequence of having gone through the rational process in an earlier life.

PIET HUT: Is there not an emphasis in Dzog Chen on holding the view even before you start practice, even if not such a long path of analysis as in Madhyamika?

DALAI LAMA: In Dzog Chen practice, in what is called the *trekchö*, or breakthrough phase, there is a breakthrough to the nature of the pristine, primordial nature of awareness, the essential nature of the mind.

In order to be able to make that breakthrough, one needs a preliminary practice. The preliminary practice entails a very careful empirical investigation into the nature of the origins of mental states, mental phenomena, and consciousness itself. It is an investigation into the manner in which they originate, the manner in which they are present—their location—and also the manner in which they dissipate and vanish. There's a very careful ontological scrutiny of these three phases of mental processes. That whole mode of investigation is completely in concordance with the logical Madhyamika view. Once that preliminary practice has been done, then you are ready to go into what is called the actual practice of identifying the nature of primordial awareness. Mipam, the great Dzog Chen master and teacher of the Nyingma tradition, said that the Prasangika Madhyamika view is an indispensable prerequisite for the actual breakthrough phase of Dzog Chen practice.

In both traditions, the close and essential relationship between rational analysis and experience, between theory and experiment, is evident. Both play crucial roles in our inquiry into the world. Theory guides experiment, and the theoretical view is the sound basis for Buddhist contemplative practice. In the view of Madhyamika, the world we are investigating is ultimately empty of intrinsic existence. The full, rich world of human experience and scientific investigation occurs within the arena of conventional reality. Conventional reality allows for the existence of even the most subtle and hidden domains. And the potential range of human capacities is—in the Buddhist view—able in principle to extend to reach them. Whatever we can validly infer about conventional reality can also become part of our experience if we suitably school our human sensibilities. Similarly, we can always refine our experiments to probe more carefully in order to confirm or falsify our scientific theories.

9

New Images of the Universe

GEORGE GREENSTEIN: When I walked into this room on Monday, I was very taken with its beauty. What a wonderful, beautiful room to have these meetings in. In the same way, the universe is very beautiful. I thought we would begin by looking at some photographs that show the beauty of the universe. Anton Zeilinger began by showing a photograph of where he lived, so I thought I'd also show a photograph of where Anton and the rest of us live.

With these words, astrophysicist George Greenstein introduced us to the grandeur of the visible universe through a set of beautiful slides of the universe, beginning with our own Earth as viewed from space. In this way we turned from the minute world of the atom to the vast expanses of the cosmos. In the process, the mysteries of the quantum were exchanged for those of Einstein's theory of relativity. Before taking up the challenges posed by recent theories of the cosmos, I will provide a brief summary of the images George walked us through on the fourth morning in Dharamsala.

Earth, when viewed from space, is a magnificent blue and white sphere, floating in a black void. It is our celestial home, a beautiful planet among several others that circle the Sun. George led us in steps from hospitable Earth to the barren, rocky Moon and beyond to the outer planets. Through images provided by the space probes Viking *and* Voyager, *we visited the red surface of Mars; giant Jupiter, with its many moons; the rings of Sat-*

Figure 9.1 A spiral galaxy (NGC3310), like our own Milky Way. Courtesy of NASA and STSci.

urn, and on past Neptune. Each image took us farther and farther from home, or perhaps better said, we extended the horizon of our imagination to include a much larger territory under that rubric. In the vast reaches of space beyond our solar system move galaxies, luminous gas clouds, supernova remnants, and galaxy clusters. These objects of deep space extend as far as we can see in all directions. The mountains of Innsbruck that Anton showed us were beautiful, but the images of these distant astronomical specimens were more beautiful still. Yet if we could look back at our own star system from some distant galaxy, the Sun is just one of a billion stars.

Picturing Our Universe

GEORGE GREENSTEIN: One of those stars is our star. It's not the biggest one, just one of those tiny, little ones. Every time I look at this photograph of a galaxy like our own [*see figure 9.1*], I get a very strong feeling of the insignificance of humanity. In the last few days, we have been speaking of how important the observer is and how important we are. I'm not so sure that's true when I look at this picture.

The final photograph was a recent one, taken by the most powerful telescope to date—the Hubbell space telescope. It shows many galaxies of various sizes and types, apparently receding into infinity.

GEORGE GREENSTEIN: This photograph is of a very small region of the sky, showing galaxies that are very distant from us. We are not sure how distant—it's very hard to know—but billions of light years distant. Every time we build a bigger telescope, we see farther, and we see more galaxies. We have never seen an end. Maybe the next telescope we build will show an end, but certainly in the past we have never seen an end.

Beyond their undeniable beauty, what can these images teach us? From them and dozens of other observational techniques, we have gained unprecedented insights into the laws that govern our cosmos and its composition and evolution. George posed the questions we all ask when we wonder at the universe.

Questioning the Cosmos

GEORGE GREENSTEIN: I want to ask very general questions about the universe. My first question is, How old is the universe? Does it have an age in the same sense, for example, that I am fifty-seven years old? If it has a definite age, then it was created. If it was not created, has it existed for an infinite length of time? How can I begin to answer this? In physics and astronomy we have ways to find out how old things are. We know the age of Earth. From radioactive dating, we can find the age of a piece of pottery, and find out how old an ancient city is. Likewise, we can date rocks, using similar techniques, and so find out how old Earth is. I could find out how old the universe is by finding the age of the oldest thing in it. I will examine everything, and I will ask how old it is. How old is Earth, and how old is Mars? We have never found anything older than a certain limit. We find a large number of things with an age up to or more or less at that limit. There are many things that are 5 billion or 10 billion or 15 billion years old. We have found many things that are roughly 15 billion years old, but we have found nothing older than that. Fifteen billion years, more or less, seems to be some kind of limit. This is not an absolutely definite limit. Perhaps tomorrow somebody will find something 30 billion years old, but it hasn't happened yet. Something interesting seems to have happened about 15 billion years ago. What was it? We don't know, but clues exist.

In 1927, Edward Hubbell discovered the phenomenon that we call the expansion of the universe. By that we mean that all galaxies, more or less, are moving away from us. I can measure how far away a nearby galaxy is, and I can measure how fast it's moving. For example, a galaxy is a certain distance away, and it's moving away from us at, say, 100 miles an hour. I can calculate how long it took to get from here, from us, to where it is now: 15 billion years. Now, let's look at a more distant galaxy. It's farther away, but as Hubbell discovered, it is moving faster, say 200 miles an hour. How long did that galaxy take to travel the distance away from us? Fifteen billion years. Every galaxy is moving in such a way that 15 billion years ago, it was here. We call that moment the big bang. What was it? Was it creation, or was it some explosion? We're not sure.

Albert Einstein's theory of relativity actually predicted this expansion of the universe. We call that version of the theory of relativity the big bang theory. Einstein's theory predicts multiple possibilities. It does not tell us which of these possibilities is true. One possibility is that the expansion will never cease: The future of the universe would be endless, without limit in time. That's very interesting because we can then start to play games and ask ourselves what the universe would look like an infinite time in the future. What will Earth be like, not a million or a billion years from now, but an infinite number of years from now? With my students, I enjoy doing calculations of what sorts of things can happen if you have an infinite amount of time for things to happen.

Another possibility is that the expansion continues, slows, stops, reverses its motion, and compresses into a second big bang, or a big crunch.

DALAI LAMA: Does the very fact that it comes together imply that there is a second big bang, or does that not necessarily follow?

GEORGE GREENSTEIN: Einstein's theory does not tell us exactly what will happen. The assumptions upon which Einstein's theory is based become false at the moment of the crunch. The theory itself becomes mathematically insane, or "singular," in the same way that one divided by zero is impossible. You cannot divide one by zero; it is illegal. It's not simply infinite; it's an error to even try. What this means is that Einstein cannot tell us what happens next. Does the big crunch turn into a second big bang? Or, alternatively, does the universe disappear? Or does some other unknown thing happen? Einstein does not tell us. Many physicists and astronomers are trying to change Einstein's theory so that it will not be singular at the big bang, and it will tell us what happens. Piet works on this, to some extent. Many people think that we will need quantum mechanics to accomplish the task, but no one knows how to do it.

I've been speaking of the future; now let me speak of the past. The present expansion must have come from a big bang. What came before the big bang? Was the big bang creation? Or did the universe exist for an infinite time before the big bang, and was it contracting for an infinite time before it hit a big crunch and then expanded into the present state?

DALAI LAMA: I recall from yesterday, when we talked about spaceships passing in empty space, there was no way to speak of absolute motion. Now it sounds like you're saying that there really is something like absolute motion. There is a real expansion, a real moving apart of galaxies.

GEORGE GREENSTEIN: It's true. It's absolutely true. In cosmology, the universe itself defines a special reference frame.

Einstein's theory gives us several possibilities: that the universe was created; that there was no creation, but an infinite past and a future disappearance of everything; or that there was no creation, an infinite past, and an infinite future. For some of these possibilities, one can actually gather empirical evidence that will decide whether such a thing happened. But for other possibilities, Einstein's theory becomes singular, and we need a better theory before we can decide. So, this is partially theoretical speculation and partially a matter of gathering evidence to see which of these is correct.

Another general question: Does the universe have an edge or is it infinitely large? I would start to answer that by gathering evidence. I would look to see how far away things are. With a small telescope I can see many, many galaxies, but beyond a certain distance the telescope cannot penetrate. So, I build a bigger telescope. . . .

DALAI LAMA: Is it not true that with the biggest telescope built thus far, you haven't come to a point where you find nothing more, or even a decrease in density of the galaxies?

GEORGE GREENSTEIN: That is correct. With the biggest telescope we have, we have seen no edge, and we have seen no decrease in galaxies. I would argue that we can conclude nothing at all from that. To discover whether the universe is infinite, we would need an infinitely large telescope.

DALAI LAMA: And the observer would also have to have an infinite life span.

A Finite Universe with No Boundaries

GEORGE GREENSTEIN: So this is a question that cannot be answered by looking directly. Now let me try a philosophical approach. Let me imagine that there is an edge of the universe. I imagine that I go out to that edge and then walk one step further. This seems to prove that there cannot be an edge of the universe, logically. But if the universe is infinite, how can I understand something that is infinitely big? Einstein found a wonderfully interesting way to escape that paradox. He found a way to imagine that the universe is actually limited but unbounded; that is, although of finite volume, it has no edge.

To determine how big the universe is, I would need to count all the locations in the universe. I ask, first, whether there is a finite number of locations or an infinite number of locations? That's what I mean when I ask whether the universe is limited or infinite. How can I count locations? Let me begin with something easier. Let me count how many people are in this room. I don't want to count the same person twice, so I'm going to take a piece of chalk and put a little mark on every person that I've counted. One, two, three, and so on. . . . This way I can make sure that I don't count the same person twice, and I see that there is a finite number of people in this room. Now I want to do the same thing with space. I imagine putting a little mark on an empty piece of space. I'll use this film box to mark the location. [*George picked up a small film box.*] Wherever I put my marker, that identifies a certain location and small volume around it. I'll make sure that I don't count the same location twice by leaving the film box there and numbering it. How big a collection of boxes can I make? Can I make an infinite pile or only a limited pile? Einstein discovered a certain mathematical theory of space, called closed space, which says there are only a limited number of places. If I build the pile bigger and bigger, eventually there is no place left for me to put the last box. The pile fills Earth; it fills the solar system; it fills the galaxy. I keep adding more and more. The pile fills all those regions that the Hubbell space telescope can observe. I keep moving, and eventually way off in the distance I see a wall of film boxes far away from me. As I keep adding more and more boxes to my pile, I keep getting pushed toward that wall. Eventually I reach that wall and find written on those boxes "one," "two," "three." . . . I'm running out of room. Eventually there is no space left, and I have used a finite number of boxes to do this. This is a universe with a finite amount of space and a finite number of objects in it. It's called closed space. A better term would be *limited space* or *finite space*. This is one of the possible mathematical models Einstein found. Other mathematical models are infinite.

ALAN WALLACE: Speaking as a mediator, this sounds like a tautology: You assume that space is finite, and then you build a mathematical model on that basis. There's no real information because the assumption determines the model.

DALAI LAMA: Is the model of a limited universe premised on a general assumption that there is no edge?

GEORGE GREENSTEIN: A few minutes ago I explained the paradox of the edge—the problem of stepping beyond the edge. Einstein's theory avoids that paradox of the edge. Let me give you an analogy. What's a Tibetan unit of measurement of area?

DALAI LAMA: Farmers measure the size of their fields in terms of how many bags of seed they use.

GEORGE GREENSTEIN: Good. How many bags of seed does it take to cover the whole planet Earth? The answer is a finite number. Nevertheless the surface of Earth has no edge. In the same fashion that the farmer does not need an infinite number of seeds to cover the whole Earth, I do not need an infinite number of things to fill the universe, in this one particular model of closed space that Einstein found.

PIET HUT: Imagine if the whole Earth were to be covered by water, a very shallow ocean only a few centimeters deep, and very flat, big fish were swimming in this ocean. They would have no idea of height. They cannot go up. They cannot go down. They only can move horizontally and their world has only two dimensions, although they can go as far as they want. They never find a limit or an edge, but a finite number of boxes can fill up their whole world. Einstein's mathematical model is similar. We are three-dimensional fish, and we cannot travel in the fourth dimension. The analogy is not completely accurate because the fish are a little bit thick in the third dimension, and our size in the fourth dimension is zero, as far as we know.

George did not want to use this analogy because the notion of the center of the sphere raises problems. Earth really has a center, but in the universe, as far as we know, there is no center. However, just as the fish in the shallow ocean will never find the center, maybe our whole universe does have a center that we cannot find because it is in the fourth dimension. In any case, this is just a model to show that you can have a finite volume without a limiting edge.

When they first encounter it, everyone is puzzled by Einstein's theory of a finite but unbounded universe. The Dalai Lama was no exception. We are all very familiar with finite volumes such as the rooms in which we sit. Such spaces are bounded by walls. In this instance walls are the limiting edges spoken about above. At every such edge, we can ask what lies beyond it. In our experience with rooms, another space always lies on the other side of the wall. Einstein's theory of relativity offers another type of finite space quite different from that of a walled enclosure. It is finite but has no walls. This possibility arises because Einstein's space-time is curved by the presence of massive bodies.

George and Piet used the analogy of Earth's surface to demonstrate the significance of curvature for two dimensions. If Earth were flat, then all finite areas (two-dimensional spaces) would be bounded by edges, as the walls dividing a farmer's fields. Without these edges, the flat plane of Earth and two-dimensional space around it would go on forever. However, another alternative exists if the flat, two-dimensional surface can bend into a sphere or other closed surface. By closing on itself, the surface area of Earth is finite but unbounded, that is, without edges.

Einstein's theory allows us to generalize this idea from two dimensions to three. In the above examples, we worked only with two-dimensional surfaces: bounded and unbounded. But our cosmos is three-dimensional. The universe might well be open and extend forever in all directions, but another possibility exists. As when we bend an infinite, flat surface into a closed but unbounded two-dimensional spherical surface, in the same way we can go from an open, infinite universe to a closed but unbounded three-dimensional universe by bending space-time. We have difficulty imagining it because doing so requires us to see the three-dimensional topology of space from the perspective of four dimensions—a skill most of us lack.

The devices of film boxes and seeds can give us a sense of how such a closed space works. Another analogy I find helpful is that of circumnavigation. Magellan sailed around the globe by heading always in (roughly) the same direction. Likewise, we can imagine setting out in any direction into the universe. No matter how long and far we travel, no matter in what direction, we would never come to a boundary, wall, or edge. Moreover, with a little clever navigation (and enough time), we could circumnavigate a closed universe.

These are all meant as aids to our intuition, which, as in quantum theory, has little experience in such matters. Of course, the mathematics of curved space is precise in all its details. Whether the universe is open or closed is not arbitrary but depends on the amount of matter in it since it is matter that curves space-time. With a sufficient amount of matter, space-time curves up into a closed but unbounded universe.

Since the 1997 Mind and Life conference, these considerations have be-

come even more complex and puzzling because of the discovery that, contrary to our expectations, distant galaxies are accelerating away from us. This has led theoretical astrophysicists to suggest a kind of negative matter, called quintessence, that pushes regular matter apart. Since quintessence entered the scene only after the conference, I will say nothing more about it.

Expansion Is the Same Everywhere

GEORGE GREENSTEIN: So, Einstein gives us two possibilities: this kind of finite universe (with no edge) or, alternatively, an infinite universe. How can we understand something that is infinite? Suppose I ask a very general question about India. I will begin by looking at a very small part of India, namely, this small portion of the tabletop right here. If I look at just this view, I see that India consists of a flat surface of green cloth. If I enlarge my view just slightly, then I say India consists of green cloth and some boxes and a glass and this piece of metal. If I look at a still bigger region—this whole room—I suddenly see that it contains people and windows and lights and so on. How big a view do I need before I get an accurately representative view of India? Maybe 10 percent or even 50 percent of India, but it is a finite amount. If the universe is infinite, how big a view do I need to represent it accurately? Say, for example, that I need 10 percent of the entire universe. That 10 percent is infinity. In order to get any representative view of the universe, I need an infinitely big telescope, and I can't do that.

If it is true that the universe is infinite, then I have no way to understand it, except for two clues. Both clues involve evidence rather than philosophical speculation or theory. The first piece of evidence is very subtle. [*George holds up an elastic cord divided into equal segments by metal rings.*] This is a model of the expansion of the universe.

Each of the metal rings represents a galaxy. George asked Alan, across the table from him, to hold the other end of the cord without moving. One ring near the end of the cord held by Alan was chosen to represent our own galaxy. George then stretched the cord as Alan Wallace held the other end stationary. Each segment expanded equally, but the speed of the expansion was greater toward George's end than Alan's.

GEORGE GREENSTEIN: This is how the universe is expanding. The galaxies very close to us are moving slowly. The galaxies farther away move more rapidly. That is what Hubbell discovered.

DALAI LAMA: Is it merely an apparent difference, or is it an actual difference in the two different speeds, close and far?

GEORGE GREENSTEIN: They really are going faster. It's not just a perception, but it is relative.

DALAI LAMA: Strange.

ANTON ZEILINGER: The strangest points come in a minute. [*At which point everyone laughed.*]

DALAI LAMA: When you taking Earth as your frame of reference, the galaxy closer to Earth is going slower and the farther one is going more rapidly. But, of course, if you shift your frame of reference . . .

George laughed and suggested that they try it. Now he held his end of the cord stationary, and Alan pulled the other end. The effect was reversed, with the opposite end of the cord stretching more quickly.

DALAI LAMA: This is simply a mode of perception—a relative perception of speed. It's not objective.

GEORGE GREENSTEIN: That is absolutely the case. Yes.

Relative versus Illusory Motion

DALAI LAMA: Now it sounds like you are confused because you said before that this is a real difference in speed, and now you are saying, no, it's only in the mode of perception. Which is it? Are you speaking about a mental phenomenon or a physical phenomenon? Otherwise it looks like our galaxy has a privileged status in the universe—that we are really at rest, all the others really are moving, and we are the center after all. Then you may as well toss in a creator, too.

GEORGE GREENSTEIN: I'm not sure what you are saying. Let me describe the situation and then ask you to analyze it. On a street in Dharamsala there is a cow that is very slowly walking. There are three taxis. One taxi is holding still. One taxi is moving slowly. One taxi is moving very, very rapidly. Each of the three taxi drivers looks at that cow and considers how rapidly the cow is moving. From the taxi that is holding still, it looks like the cow is moving at a certain speed. Maybe the second taxi is driving right along with the cow. What will that taxi driver say about the cow?

ALAN WALLACE: You are trying to say that it relates to the frame of reference and not the mode of perception.

GEORGE GREENSTEIN: What is the difference?

ALAN WALLACE: If it's a frame of reference, then you are dealing with the physical reality. If it's merely a perception, it's like an optical illusion. Are we talking about an optical illusion, or is it really a matter of the structure of the physical universe?

GEORGE GREENSTEIN: What we're talking about is how we measure velocity. Suppose a bird flies through this room right now, and we all see that it's going at a certain speed. Is that an illusion or a perception of truth?

ALAN WALLACE: It's a perception of what happened.

GEORGE GREENSTEIN: Why do you raise the possibility that it might be an illusion?

DALAI LAMA: If we denied the truth of the event, then we would have no basis to talk about any valid perception of it.

GEORGE GREENSTEIN: But can we all agree how fast the bird was moving? Suppose that I was walking along in the same direction as the bird when it flew by. Will I judge the bird's speed the same as you will?

DALAI LAMA: Then it's just a matter of relative speed. It's not an optical illusion; it's the frame of reference.

GEORGE GREENSTEIN: Well, that is exactly what we mean when we talk about the velocity measured through telescopes of distant galaxies. I am fascinated because I cannot understand why this is difficult to communicate. It's a clue that we are exploring something very interesting in the differences between our understanding.

DALAI LAMA: The problem arises because this is not a matter of perception. The speed at which they are moving can be measured and so seems like an objective reality.

GEORGE GREENSTEIN: It's a relative speed.

THUBTEN JINPA: But it can be measured.

GEORGE GREENSTEIN: In the same way, we could each measure a different speed of that bird flying through the room. There is no absolute velocity.

DAVID FINKELSTEIN: When we say velocity we always mean relative velocity. There is no such thing as absolute velocity in physics.

Big Bang as a Centerless Explosion

We naively imagine three types of velocity perceptions. A blow to the head may set the world spinning, but we know this is merely an illusion, or a mental phenomenon *in Buddhist terminology. The physical phenomena of motion are really all of one type, namely, relative motion. We quite commonly slip, however, into assuming that there is a true or actual physical velocity for all objects. This is especially tempting in the case of the big bang theory. We are tempted to say that the expansion of the universe must be away from some center, as when a firecracker explodes, in which case the hot gases and bits of paper fly from the original location of the fire-cracker. We are convinced that this is not a correct picture for what oc-curred at the origin of the universe. Our best current theory describes a centerless expansion. This is a very difficult concept, and an analogy may help.*

Imagine a balloon with small spots. As we blow up the balloon, the spots are farther apart from each other. No one spot on the balloon is in a privileged position. None is the center around which the expansion occurs. Or, one can also say, all spots have equal claim to be the center. That is, if we were to view things from any one spot on the surface of the balloon, we would see all the others spots receding away from us. This is exactly what astronomers do see. The balloon's surface is a two-dimensional, un-bounded but finite area. The universe has three spatial dimensions. We be-lieve its expansion is entirely analogous to that of the balloon. From Earth, the distant stars and galaxies are seen to be moving away from us. How-ever, from the vantage point of some other galaxy, the whole universe will be seen as expanding away from it.

The expansion of the universe is not an illusion, but different observers located at different places in the universe will, of course, see the expansion from their own vantage point. The only consistent account we have dis-covered is one in which everyone sees the universe expanding in a similar way. These ideas proved quite difficult to convey. George Greenstein's cen-tral point was the one he ended with. Of all the different ways in which the universe could be expanding, only one model can offer the coherence and consistency we seek, and it is this one.

GEORGE GREENSTEIN: I thought this would be an easy means to an end, but it wasn't. The end I am seeking is to understand infinity. But let me reach my point. Living on Earth, we see that the more distant a galaxy is, the faster it is moving. An alien being that lives on a distant galaxy also sees that the more distant a galaxy is, the faster it is moving. The universe is expanding in such a manner that every observer sees the same thing, and there is only one possible mathematical model for this. Mathematically speaking, there is a large number of different ways the universe could be expanding, but in all those other ways, different ob-servers would see different things. And because every observer's view of the universe is the same as every other observer's view of the universe, this means that you do not have to look infinitely far to understand an infinite universe. You can look locally, and that is good enough.

DALAI LAMA: This does not prove that the universe has no limit, but it does mean that you can make a judgment. Are you saying that in order even to conceive the possibility of an infinite universe, you don't need an infinite measurement?

GEORGE GREENSTEIN: Yes. But had the universe been expanding differently, we would have needed an infinite measurement.

I don't know what to conclude from this argument. I find it very striking that the universe happens to be doing something that makes it possible for us to think about it. It could have been expanding differently, in which case we would have perhaps found it impossible to study empirically.

To summarize, one of Einstein's various possible models of the universe is infinitely oscillating—a big bang, expansion, a big crunch, expansion, repeating endlessly. Another model comprises an infinite past and an infinite future. But these models raise questions that we find very hard to think about. How can we understand creation? In many of Einstein's models, it is possible that the universe was created in the big bang. Was time created? Was space created? Did the laws of physics exist prior to creation? What does it mean to say that the laws of physics exist, but the universe doesn't exist? These are the questions that I find impossible to think about. Another question: In the view of the universe that we see, people seem to be irrelevant. All of life seems to be irrelevant. How can we amalgamate this view with the view we've been discussing in which the observer is all-important?

In this brief rendering of the story of the universe, large questions remain unanswered. They probe the foundations of our concept of reality, of space, time, natural law, and the conscious mind. Some of them are the focus for the next session's discussions.

10

Origins of the Universe
and Buddhist Causality

Our two astrophysicists, George Greenstein and Piet Hut, continued their description of the evolution of the universe. One difficult point in cosmology is the proper interpretation of telescopic observations. With our largest telescopes, we can see very deep into space, but we must bear in mind that in doing so we are simultaneously seeing back in time. The light that reaches our eyes has traveled for a long time—up to several billion years. We can only infer the current state of affairs in the distant reaches of the universe because our direct astronomical observations are always of events long past. However, from a careful analysis of these observations, we can construct a likely story of how our galaxy and, indeed, the whole universe formed and what it is like even in those sectors currently unseen by us.

PIET HUT: The present distribution and configuration of stars in the galaxy is an imprint that results from how the whole galaxy was formed. Originally, a gas cloud contracted, and while contracting it started to rotate. The rotation became faster, and then everywhere stars were formed. The stars still have the same general motion as the original gas. They inherit the rotation of the gas, and because the gas was denser in the center, more stars were formed there. There are other galaxies where the formation was more complicated and they don't rotate. They are not flat, but they can have an arbitrary shape. They may be round like a ball or a more elongated shape like a rugby ball. The movement of stars within a galaxy like that can be completely random. Our galaxy happens to be moving more systematically.

GEORGE GREENSTEIN: And then there are clusters of galaxies. These clusters do not have the same pattern of movement as the solar system. Their galaxies move in complicated paths, with no general tendency for rotation. Clusters seem to be random. They look like clouds of mosquitoes, round or irregular.

Finally, these clusters seem to form into a very large pattern, only recently discovered, that looks like a froth of soap bubbles. The galaxies lie in curving sheets that intersect like the surface of soap suds, and there are vast regions like bubbles with nothing in them, called voids. These are the biggest structures we've ever seen, but if we build bigger telescopes, perhaps we will find bigger structures.

PIET HUT: However, we expect that there will be no new structures even if we see much further. We expect everything becomes more or less the same. The reason is that if you look very far away in space, you also look back very far in time. You can look so far that you reach a distance or time before galaxies were born, and the only thing you see is the glowing light of the big bang. That light is very regular, almost the same in every direction. There are no soap structures, no patches. The difference in the intensity of the radiation is less than one-hundredths of 1 percent if you look in different directions at the sky.

It is interesting that we still don't know whether the universe will keep expanding or will collapse again. But by looking at the afterglow, we hope to get the answer in five or ten years. Very often in astronomy, what is far away is simpler than what is nearby. Even in our solar system, Earth is much more complicated than the Sun. If you dig into Earth, you find different rocks in different places. The Sun, which is a ball of glowing gas, is much easier to understand. We know much more about the center of the Sun than the center of Earth. Similarly, the further back we go in time, the more we know because the universe was simpler then. The glowing background radiation is much simpler than the galaxies, which were formed later.

DALAI LAMA: I asked earlier whether you found any limit or even a tapering off of the density of galaxies as you are look very far away, and George said no, which implies that they could go indefinitely, as far as we know. But now you are saying that if you look so far, then you find an afterglow beyond the galaxies, which sounds like you're reaching an actual limit of the physical mass of the universe.

PIET HUT: We have no reason to believe that there are fewer galaxies in space beyond a certain distance, but we cannot look only in space. We have to look in space and time. The farther we look in space, the earlier we look in time; and indeed, if you look far away in time, the galaxies disappear. At the time of the afterglow, there were no galaxies. The galaxies were born probably when the universe was a few billion years old—we don't know exactly how old. The next generation of telescopes probably will show us the birth of galaxies. We have made a few observations of what look like "child galaxies" just being born, but our telescopes are not quite good enough.

ARTHUR ZAJONC: Let me try to explain this difference between space and time. We imagine that the universe started over a very large region of space, rather than starting at a single point. There was a creation—a big bang—but almost instantly the universe was huge, 100 billion light years in extent. Then the whole thing, 100 billion light years across, evolves in time. We are now 12 billion years away in time from that creation. There's no way we can see to the edge of the space of the universe because the light from the 50 billion light years away has not had time to reach us. So, the horizon we see is not a horizon in space; it's a time horizon. We simply can't see that far. We haven't been around long enough to see out to the galaxies that may exist at 50 billion light years away.

DALAI LAMA: But if you're suggesting that at the big bang, the universe could immediately have been 100 billion light years in diameter, then the whole notion of expansion of space gets lost.

PIET HUT: No, no, no. What happened is that in the big bang, and this is a point which is often misunderstood in popular astronomy books, different points were going away from each other much faster than the speed of light. It sounds like a contradiction because Einstein tells us that the speed of light is the highest speed possible. It's true that if two objects pass each other, their relative speeds cannot be more than the speed of light. But in the explosion of the big bang, the speed between different points can be much more than the speed of light. You can even have an infinite universe, which is simultaneously exploding. At the present time, we have a finite time depth, but we may have an infinite space.

We know that time is finite since the big bang, but we don't know whether space is finite or not. The new observations of the slight differences in the afterglow will tell us whether the universe is finite or not. The afterglow was discovered thirty years ago, and for twenty-five years people have searched for these small differences in temperature in the afterglow. They finally found them, about five years ago, after twenty-five years of searching. Now that they have found them, they can build instruments to get more accurate measurements, and so in five or ten years they will know how much matter there is. From these small differences in temperature, the galaxies were later born. If you look at the properties of these seeds, you can know more about the properties of the galaxies, and of the whole universe, including whether or not it will collapse.

DALAI LAMA: Does the big bang theory include any kind of postulation of where it really began, in relation to our galaxy?

GEORGE GREENSTEIN: It was here.

PIET HUT: It was everywhere. It was a centerless explosion.

ARTHUR ZAJONC: You thought quantum mechanics and the superposition principle were confusing. Now you know that cosmology is even more confusing.

As everyone who first hears these ideas, the Dalai Lama was having difficulty conceiving of a centerless explosion, of an expansion that takes place everywhere at once. In addition it is difficult to remember that observation into space is also observation back in time. For example, consider an event 100 light years away, which took place 200 years in the past. If a supernova took place at that point in space and time, we could see its light because it takes 100 years to reach us; but since the supernova was in existence for 200 years, there is plenty of time for the light to reach us: It would have arrived 100 years ago. Now consider a second supernova that occurred 200 years ago but today is located much further away from us, say, at 500 light years distance. We could not yet see the second supernova event because its light will take 500 years to reach us. In another 300 years, astronomers will notice it in the sky.

In other words, there are many events in the universe that we cannot yet see because the light signal from them has not yet reached us. This is true for the very earliest events also. Suppose the universe is 12 billion years old. If shortly after its birth some event like the formation of a galaxy took place 50 billion light years away from us, we could not see that event now. We would have to wait about 38 billion years.

Piet described this situation. Astronomers look out in space and simultaneously back in time. We see galaxies 10 billion or so light years away and know, therefore, that they have existed for the last 10 billion years. Galaxies may also exist 50 billion light years distant from us, but we cannot see them because their light has not reached us yet. It is conceivable that galaxies could go on forever in an infinite universe; we cannot tell yet whether they do or not. What we can say with some certainty is that a big bang took place roughly 12 to 13 billion years ago. If we look back in time beyond 10 billion years, we see only child galaxies and then the cosmic glow at three degrees above absolute zero, which we take to be the remnant of the big bang.

DALAI LAMA: With a universe that is finite in time, is it possible for there to be an infinite number of galaxies?

PIET HUT: Yes. But there are only a finite number of visible galaxies.

DALAI LAMA: This is very problematic. You have the beginning of the entire universe at a finite point of time, and expansion starts from that point. If you have a finite duration of time, unless galaxies are proliferating at an infinite rate within any duration of finite time, you will end up with a finite number of galaxies.

PIET HUT: Even if the universe is finite, the same type of problem exists because the explosion happens so fast that different parts of the explosion have no causal communication, no time to propagate signals to each other. If I look in one part of the sky, that portion of the afterglow has had no communication with another part of the afterglow. The communication length at the time in which the currently visible afterglow was produced was a small angle of a few degrees on the sky. The uniformity of afterglow is a puzzle because there could be no causal connection. How did the explosion start the same way in all places? It was not like an explosion, where one thing is touching something else. Nothing touched; not even light could touch. If you can have an explosion in a finite universe, without the different parts touching, then you can have an infinite explosion in which the parts do not touch and still all move in the same way. Infinite or finite, we have the same very important problem. We probably need quantum mechanics to solve this.

The Possibility of an Oscillating Universe

DALAI LAMA: So far we have had no discussion about the possibility of whether there was one big bang or several big bangs in sequence. Also, there is no reason in principle to deny the possibility of whole other universes with their own big bangs, with which we have no physical contact.

GEORGE GREENSTEIN: This is true.

DALAI LAMA: If that is a possibility, then we could have a different angle on this problem of infinity. What is the mainstream position on the question of multiple big bangs?

ARTHUR ZAJONC: We can distinguish two parts to your question: a sequence of big bangs and the possibility of very distant big bangs, perhaps simultaneous, perhaps not. Let's first address the evidence for an oscillating universe with sequential big bangs. This relates to the question of dark matter and whether the universe will expand forever or whether it will expand up to a certain point and then collapse. George, could you explain why we don't know?

GEORGE GREENSTEIN: Einstein tells us what sort of evidence would answer the question of whether the universe will continue expanding or collapse and then expand again. It depends on how much matter there is in the universe. If there is more than a certain critical amount of matter, the universe will oscillate. If there is less than that amount of matter, then a single expansion will persist forever. If we try to measure how much matter there is in the universe, we get embroiled in a fascinating complicating factor: There seems to be much, much more matter than we can see with our telescopes. We detect this dark matter through its gravitational attraction. Newton teaches us that matter exerts a force of gravity, attracting other things to it. We can detect how much gravitation there is in our region of the universe, and there is much more gravitation than you can account for with all the stars, all the galaxies, all the planets.

Within one or two or three light years from us, there is maybe two or three times more matter that we cannot see than matter we can see. If we consider our entire galaxy, there is maybe ten times more unseen matter than seen matter. If we consider clusters of galaxies, there may be ten, twenty, thirty times more unseen matter than seen matter. It appears that we have never noticed most of the universe. We don't know what this dark matter is. I find this ominous.

The question is whether we now have evidence of enough dark matter to make the universe oscillate. The answer is no, not quite. But there is so much uncertainty as to how much dark matter there is, that there may possibly be enough. This is one of the most exciting areas of research today.

ARTHUR ZAJONC: What Piet was describing is an attempt to solve this question by another means. By measuring the afterglow of the background radiation very precisely, astrophysicists hope to reveal the answer to this question.

PIET HUT: And, of course, if the universe collapses, we don't know whether it will expand again or not. It would seem a little bit easier to start again if it collapses, but we don't know the theory. You mentioned multiple universes. Indeed, there are some speculations that after the big bang, different places in the universe could be the seed for a new big bang. A black hole in our universe might lead to another universe being born. We don't know, but this theory of the "multiverse," as opposed to the universe, is a possibility.

ARTHUR ZAJONC: This is a very, very exciting topic; but we'll have to come back in fifteen years to tell you the answer.

ANTON ZEILINGER: That has been said very often in the history of science: Come back in fifteen years. And the answer did not come; the problem just sounded more complicated. I remember people saying, "Give me one piece of the moon and I will tell you the history of the universe." It did not happen that way. We got one piece of the moon, but it turned out to be more complicated.

What Caused the Big Bang?

DALAI LAMA: If, for the sake of argument, we assume that a singular big bang is the more logically consistent position, how would one account for the origin of the big bang?

PIET HUT: At the moment, we really cannot say. We have a rule in physics that if you predict a singularity or something infinite, whether infinitely small or infinitely anything, it means that your theory has broken down and you need a new theory. We don't believe that the universe was really infinitely small; we believe that because the theory predicts it was infinitely small, we have to invent a new theory. People are working on that. It may take ten years, maybe a hundred years; we have no idea. But we are hopeful because we are still making progress in our theory. Hopefully the progress will continue, and hopefully we will have an answer sometime.

GEORGE GREENSTEIN: What is the Buddhist position about the cause of the origin of things?

DALAI LAMA: Looking just at the physical universe—because that's not all that Buddhism counts, as you know—there is a stream of substantial causes, which means things turning into other things. If you trace developments back in time, you always have something out of which the present stuff arose: This arose from previous stuff, which arose from previous stuff, which in turn arose from previous stuff. The form changes. The manifestations and configurations all change, perhaps drastically, but there is a kind of conservation principle in action. You never have nothing turning into something. If you posit no beginning, then you can simply trace back indefinitely. If you do posit an absolute beginning, then that would be something absolutely without cause.

GEORGE GREENSTEIN: Is causation a concept that only applies to things after creation?

DALAI LAMA: Buddhism does not posit an absolute beginning. A causeless absolute beginning flies in the face of Buddhist logic. If we reject that as untenable, then any specific cosmos or world system coming into formation would be derived from the residue of the preceding universe, which went through a destruction period. This is where the whole Buddhist theory of space particles comes in. When the universe is destroyed and dissolves, all of the other elements dissolve back into space particles. From these space particles, a catalyst will strike, and the formation of the next world cycle begins. But there is a continuity, a preservation. This notion of space particles is not a pan-Buddhist assertion. It is probably unique to the Kalachakra system, but that system is given a very high profile within Tibetan Buddhism.

ANTON ZEILINGER: How about the possibility, Your Holiness, that space and time came into existence together with the universe? In this picture, the whole question of time before the big bang is not applicable. If time begins at that moment, then there is no before, and therefore the question of cause and effect is empty.

DALAI LAMA: The problem here is that, just as Buddhism does not posit any absolute beginning point for the creation of the universe, similarly Buddhism would not posit a finite beginning of time or of space. One can talk about the beginning and end of a particular world system but not the universe as a whole.

Free Will within a Causal World

The discussion continues informally through the tea break.

PIET HUT: How about free will for human beings? Do we have free will or is every action caused by something?

DALAI LAMA: What exactly do you mean by free will?

PIET HUT: That I am responsible for my actions. If my actions are caused by something in the past, how can I be held responsible?

DALAI LAMA: Generally speaking, Buddhism would accept that human beings do have free will. Many of their actions are determined by the individuals themselves. Buddhism does not posit a creator. Of course, Buddhism does talk about karmic imprints being carried on from one lifetime to another. But at the same time, Buddhism allows for the potential of such karmic imprints to be neutralized or increased. It is an ongoing process.

PIET HUT: If there is room for personal action and personal choice, then things are not completely fixed. Why do you think that the universe is more fixed than human beings?

DALAI LAMA: Because it is material. But even in the material realm, Buddhism would not say that the universe is deterministic. The Buddhists only posit that it must have a causal continuum. Let us take the example of someone taking the train to Delhi from Patankot. You may have made up your mind and bought tickets, and all of that, but until you have actually boarded the train, it is not determined. There's always room for change or fluctuation. Similarly, in the case of the formation of the universal system, until the actual evolution begins, one could say that it's not determined.

ALAN WALLACE: It's like a superposition state.

DALAI LAMA: From the Buddhist point of view, the karma of all the sentient beings that inhabit the universe plays a role in shaping the formation of the universe. Once the actual physical evolution begins, then there is a determined path.

ALAN WALLACE: At that point it's like the collapse of a wave function, and the path is inevitable. But there's a phase of uncertainty before the fully ripened effect of karma sets in. Then there is a shift, the uncertainty period stops, and a deterministic phase begins for a particular sequence of events.

ANTON ZEILINGER: Can I ask a very general question? Where does this reasoning come from that you do not accept that things could happen without cause? Is it related to the understanding that otherwise we would have to accept the existence of God, or are these two things completely unconnected?

DALAI LAMA: It has nothing to do with God. The Buddhist position is not a theological argument at all. It's a purely philosophical and logical argument. If you posit an effect or an event with no contributing circumstances that give rise to it, then either it should be happening all the time or it should never happen. There's no way to account for its occasional nature.

DAVID FINKELSTEIN: But causeless things are happening all the time.

DALAI LAMA: We don't know that. That's what's being debated here.

ANTON ZEILINGER: But isn't that a contradiction in itself? If you require that something acausal should happen all the time, then you assign it a structure that goes beyond acausality. An acausal thing should be signified by the fact that it does not happen all the time, but at times we don't have any reason to expect it.

DALAI LAMA: I think that probably there is a problem of semantics here because when Buddhists use the term *cause*, it has a very broad meaning. In the Western context, *cause* seems almost identical to explanatory cause: something that could account for a result. When Buddhists argue that no event can come without a cause, they are not saying that every event can be accounted for and explained.

Of Mind, Body, and Karma

ARTHUR ZAJONC: Could you describe this a bit more? When we were talking about the randomness of quantum events, you considered the possibility of karmic conditions affecting the outcome. But then you said, no, this doesn't apply. Are there categories or levels of causation beyond strictly material causation?

DALAI LAMA: There are three whole classifications of causes. There are physical causes. There are purely mental causes, which do not consist of mass or energy. Finally, there are nonassociated composites, things like time, which are neither physical nor mental structures. Another classic example of a nonassociated composite is an individual, a person. You are not a physical phenomenon. Arthur Zajonc is not composed of mass and energy; only your body is. And you are not simply a mental phenomenon. You are a person who has a body and a mind, but you are not a body and you are not a mind.

ARTHUR ZAJONC: Would it be possible to account for hidden variables with mental causes or nonassociated composites? Perhaps, down to the quantum level, material causes operate in the normal way. But it would still be an open question as to whether these other two types of causes might contribute to what appears to us to be random.

DALAI LAMA: There definitely could be cognitive influences there. That would bring us into the discourse of karmic relationships to these events, which principally relates to the resultant effects of pleasure and pain. That is the chief issue within karma.

ARTHUR ZAJONC: Some people would say, "If the first level of physical causation accounted for everything, then there would be no opportunity for the other two." So, the randomness that appears when viewed from the first level is also an opportunity for the inflow of the other two levels of causation. This is the classic mind-body problem. How is it that the mind could affect the body? If the body is constrained completely by material causes, then there is no room for the mind. If, however, there is a limit beyond which the physical is apparently random, then you have the possibility of mental causes.

DALAI LAMA: I think that here we have to be quite clear because the light that is registered on the detector has a cause. It is very simply the light coming from the source. That is fairly straightforward. The noncausality that is referred to in the context of quantum phenomena is the randomness of the photon hitting the detector. So it's a slightly different question because we are looking for the cause of an event rather than a thing. I don't think it is very likely that there are nonphysical mental causes that are acting as hidden variables in the double-slit experiment, for example, or in radioactive decay. It just simply goes on. Radioactive decay takes place no matter whether anybody is watching. Neither consciousness nor karma is in any way relevant here.

Then we raise the issue of quantum events within the brain. The brain definitely is related to consciousness, and there are quantum events taking place in the brain. Must we necessarily assume that all of the causal influences taking place in brain events are purely of a physical nature and none of them of a really mental nature? There's no reason to think that. When you're dealing with the brain, there may indeed be mental causation. Habitual propensities or karmic imprints that are in one's mindstream may also be an influence. So that may be much more complex.

ARTHUR ZAJONC: Very interesting. Very controversial.

Later that day and the next morning, a few of us picked up this discussion informally. Only parts of the conversation were taped, but the gist of it is as follows.

Most physicists today believe in the fundamental statistical character of quantum mechanics. Nature is simply made this way. The idea of objective randomness maintains that even if we knew all material causes affecting a particular event, at some subtle quantum level the event would still be uncertain. Others—including a few prominent physicists such as Einstein— have protested. Objective randomness has not been proven, and there may well be a more subtle causal account that is completely consistent with conventional quantum mechanics but gives more detail about the factors behind each event. David Bohm's so-called ontological interpretation of quantum mechanics offers such an account, for example.

If we tentatively accept the fundamental statistical character of quantum mechanics, we might argue that every event in nature, whether it is the roll of the dice or the falling of a raindrop, possesses a minute residual de-

gree of objective randomness. Everything is quantum mechanical at some level. The objective randomness of quantum mechanics then becomes a kind of window for nonmaterial causes. What is the character of these subtle immaterial causes? Do they correspond to what Buddhism would call mental causes, or karmic disposition?

The argument goes, There is a limit to the level at which physical principles can apply. Beyond this limit ambiguities arise. Into these ambiguities other factors can work, such as those mentioned in Buddhism. We can skew events by our intentions or will, but in a way that does not violate the laws of physics. This is one way to approach the mind-body problem. Another way to resolve it is to eliminate the mind (or the body) altogether or somehow to deny it any ontological standing whatsoever. Clearly, Buddhism rejects such a reduction. In Buddhists' worldview, mind has as much standing as body.

The question still remains, How can extremely small effects of quantum mechanics make a difference at the macroscopic level? Into the chaos of mechanical causation might come the minute perturbations of mind, but it would seem that these subtle prompts could not make any difference. This is true enough. However, the study of nonlinear dynamical systems shows that under certain circumstances small influences can be amplified dramatically, even exponentially. This is called sensitive dependence on initial conditions, or the butterfly effect.

Thus, some speculate that a combination of quantum uncertainty and chaos dynamics could provide the basis for the mind's effect on the body. In this way, mental factors might enter the physical organization of sentient beings like us.

Conceptions of Causality

David Finkelstein brought us back to the issue of causality. He distinguishes between causality in Einstein's sense and determinism. Einstein's relativity theory places a simple but powerful constraint on causality, namely, that causes cannot propagate faster than the speed of light. Thus an event at one location cannot "cause" something to happen at a distant location instantaneously. Rather, some kind of disturbance or signal must travel from one location to the other, and the fastest it can travel is at the speed of light. Notice that this constraint says nothing about the mechanics of causality; rather, it is a simple but universal constraint. Determinism, by contrast, gives us assurance that the detailed time evolution of a system under study is completely determined. One can imagine a breakdown of *determinism at some level, but without violating Einstein's causal constraint. Quantum mechanics is an example. Whereas the wave function evolves deterministically, according to the time-dependent Schroedinger equation, the individual events of the physical world are not deterministically bound. With this as background, we can return to the conversation.*

DAVID FINKELSTEIN: I wonder if there is an error in communication. You spoke of a causality that is not deterministic, and I don't think that would cause any difficulty as a description of what we see in quantum mechanics. If not a breakdown in causality in Einstein's sense, it's a breakdown in determinism. It would in fact be very hard for me to understand how a meditator could come to the conclusion that the same beginning always leads to the same end. I imagine one experiences things just happening.

DALAI LAMA: This is still an area of confusion. To begin with, Buddhists have never really applied the analysis of causality at the level of photons and electrons, so it's very difficult to try to fit quantum physics into a wider Buddhist framework. Also, Buddhism often talks about continued existence, or a continuum, in relation to causality. This does not mean that the same substance continues all the way through in the sense that fire has to burn all the time to be considered a continuum of fire. Sometimes a continuum can be understood in terms of a potential. For example, in the case of a continuum of consciousness or a cognitive event, the continuum is understood as some kind of imprint or disposition. Similarly, sometimes the continuum of a material object is understood in terms of a potential, so it need not necessarily be actualized. For example, when we talk about a future event or a future object, the Prasangika school defines this as something that exists potentially but whose conditions haven't fully formed.

ARTHUR ZAJONC: You might be interested in Aristotle's definition of light. Let's imagine this room has no windows, so there's no external light, only the ceiling lights. If I turn all of these lights off, I see nothing. Aristotle would say that the room is potentially transparent. If I turned the lights on, the room becomes actually transparent. Light, for Aristotle, is the actualization of the potentially transparent. Some of the thoughts of Buddhism are not too dissimilar.

TU WEIMING: My very amateur reading of how the classical notion of causality becomes problematical opens up new possibilities for dependent origination or dependent arising. And if you look at that, not from the point of view of physics, but in sociology or many other areas, it becomes very significant. One example: The great sociologist Max Weber was very much intrigued by the relationship between the Protestant ethic and the spirit of capitalism. He came up with the term "elective affinity" to describe the relationship, which is not causal. Things happen concurrently but not causally. There are patterns, sometimes showing correspondence, sometimes reminiscent of objective randomness. In this case, you can't posit causality in the classical sense. You recognize there is something there that should be explained or understood. These phenomena indicate a confluence of some mutually influenced forces. Of course, there may be hidden causes and many, many other kinds of reasons. The Buddhists are remarkable in trying to understand the multidimensional nature of the various kinds of relationships. Of course, there are hidden causes. Of course, there are motivations. It's not conjecture of the mind. It's not just optical illusions. So, my sense is that this opens all kinds of other possibilities because the quantum view itself has its own self-reflexivity.

DALAI LAMA: Perhaps the confusion arises from the different contexts in which the Buddhist discourse on causality is taking place. The classical discourse on Buddhist causality is presented by the realist schools. However, in Nagarjuna's *Mulamadhyamikararikas,* the fundamental text for the Middle Way (Madhyamika Prasangika) school of philosophy, which for Tibetans represents the apex of Buddhist philosophical thinking, there is a discussion of simultaneous causation. Nagarjuna describes one form of causation as the causation of mutual dependence. He talks about causal dependence, which is simultaneous and mutual, between the actor and the act. In the classical sense, dependence was only one-directional. The effect depends upon the cause, but not the other way around, whereas the Madhyamika-Prasangika school speaks of a mutual causal dependence.

The Future of Science

From these rather detailed musings about causality, we returned to broader themes, looking ahead with David to the next stage of scientific challenge and discovery.

DAVID FINKELSTEIN: In order to go beyond the big bang, we have to go beyond our present physical lives. Since the big bang is supposed to be a point singularity in the present theories, it's also clear that quantum effects will be very important in the actual situation. If there is a quantum of space and time, for example, the question of a point singularity doesn't arise. The problem is to find it.

People have been looking for the next physical law for fifty years, and there are still only traces of it. It's clear we're at the foot of another plateau. There's probably another access to relativity coming. I've tried to make a list of all the previous approaches to relativity to get a running start for the jump to the next plateau. I see that, roughly speaking, the East began about 1,000 years ahead of the West in this race. For example, the relativity of position, or the plurality of worlds, was first recognized in the West in the time of Bruno, around 1600, and it was already over 1,000 years old in the East. We began to catch up a little bit with the relativity of velocity, which I think was understood first in the West, and the relativity of time associated with Einstein. With quantum theory, the East was again far ahead of the West in an important sense. This is because Western cosmology began with planets and worked down, and was astonished to find how sensitive electrons and photons are, whereas Eastern cosmology began with thoughts and extended from there. It's not a bit surprising that electrons are sensitive. Thoughts are even more sensitive.

DALAI LAMA: What do you mean by sensitive?

DAVID FINKELSTEIN: They respond to observation or are changed by observation. The essential point of quantum theory really is that every observation necessarily changes what is observed. This is obvious to every meditator.

The next question is the law. Here, I'm really surprised by the position that the West is still taking after all these years. It's again a situation where we look for something that acts without being acted upon. We inherit the idea of the law from Einstein. It's an equation, written down in a book, which is completely known and influences what happens in the world. There's no counterinfluence. What happens in the world does not affect the law. This one-way action up till now has always been the sign of a degenerate theory beyond which a more symmetric action lies. It seems very natural now to consider at least the possibility that the law, too, evolves and is subject to change by what happens in the world. Since the world is quantum, it seems inevitable that the law, too, has a quantum nature, and this means that it can never be completely known. It can never be written down.

This is a very difficult thing for a Westerner to accept. I feel a little embarrassed now at having spent so many years looking for something that could be completely known. It's a little undignified. I like the idea of spending the rest of my years looking for something that cannot be completely known. I must say that I'm not being at all original in what I'm doing here. I'm never original. I'm just doing at the quantum level what Einstein did at the level of geometry. Einstein said geometry affects matter; therefore, matter must affect geometry. The laws of nature affect matter; therefore, matter must affect the laws of nature. The next step Einstein took was to say that if the geometry is a physical object, who needs anything else? Maybe geometry is all. This is called unified field theory. It didn't work. The corresponding thing at this level would be to say that if the law is an evolving physical entity, a quantum entity, who needs anything else? This would be a very difficult question in the West. I wonder how difficult it is for the East?

DALAI LAMA: If one asserts the existence of these natural laws, then from the Buddhist perspective would there be any need to posit the existence of anything else—space, time, knowledge, matter? Probably not. Those laws would be sufficient. If you compare modern Western physics and Buddhist physics, in Buddhism a lot of the discussion of natural laws is presented in a wider context that would cover all three realms of phenomena: matter, consciousness, and abstract composite entities. Natural laws are presented as something that would cover the features common to these three realms. But there is not much minute discussion of physical laws found in the classical texts.

THUBTEN JINPA: Mainly because the Buddhist scholars did not have access to telescopes or microscopes. They had only thought and experience to come up with the theories about the physical universe.

Toward a Cocreative Image of the Human Being

TU WEIMING: At this juncture I'm very pleased to share some reflections inspired by conversation with David.

We've been treated with a glimpse of the smallest and the biggest, all that current scientific investigation and informed imagination are capable of grasping. The vital role of the observer strongly suggests a new vision of the human person in this whole enterprise. With the emergence of new physics and cosmology, many of the social and cultural values that the scientific revolution has contributed as part of the enlightenment project of the modern West are now outmoded, or at least problematical. The idea of progress through history, from religion to metaphysics to science, then in the development of science, has left behind many old forms of knowledge as irrelevant or superseded. That is not right, and our discussion seems to show this clearly. We need to seriously reexamine whether rationality—rather than aspects of human faculty such as sympathy and compassion—is by far the most important area of human development. I think that faith in the reductionist model is also outmoded, hoping to reduce complex structures to the extremely simple if we analyze something deeply enough.

There are many forms of ancient wisdom in the major spiritual traditions, and Buddhism is outstanding here. But even many of the indigenous religious traditions, such as the Hawaiian, Maori, and Native American, seem to hold some enduring values for us. Particularly significant for us is the shared concern for the knower, the observer—in other words, the person, and especially the person's self-knowledge and self-cultivation through many different forms of spiritual exercises. What does this mean?

First of all and at a minimum, the reductionist view of the human person has to go. A human being is not simply a rational animal. A human being is not simply a tool user. A human being is not simply a linguistic being. A human being is poetic and aesthetic, is capable of sensitive responses to an ever-expanding network of relationships within the human world and beyond, even with distant stars. Consider how we are awed and overwhelmed by the pictures of the universe and by the experiments. Human beings are social beings, with an emphasis on relationships and connectedness. Human beings are political; judicial but also historical and religious, searching for ultimate meaning; and philosophical, involved in continuous processes of self-reflection. Human beings are not merely observers, or even experimenters, but also actors in the process—actors in this sense of cocreators.

There is a partnership here between humanity and nature and also, for the religionists, though not necessarily the scientists, a mutuality between humanity and heaven. Human beings are cocreators in two senses. One is a weak sense: In the process of developing our knowledge, we are never totally disinterested, dispassionate observers, trying to discover something real out there. That is the wrong image. We are actively involved in that process: in the construction of instruments, in the positions we take, or in the time and place that we happen to be because of karma. We know that our contribution in the process of acquisition of knowledge has to be calculated, even if we consider ourselves totally insignificant from the perspective of the stars and galaxies. It is a joint venture with reality, and our whole being is being involved.

The strong sense in which we are cocreators is more difficult. I offer this with a sense of awe and humility; if it is misunderstood, it would be blasphemous from a religious point of view. The old humanism that emerged out of the European Enlightenment and scientific discovery was very much a human-centered ideology. It was anthropocentrism as a form of scientism and not science, but scientism, as an ideology. It involves a total rejection of spirituality and a very aggressive approach toward nature. It's neither naturalism nor spiritualism; it's humanism based upon the idea of knowledge as power. In this context, the human claim to be a cocreator rests in the belief that the secret code of the universe is implanted in the very nature of being human. The attempt to decode this particular secret message is, in principle, humanly possible through self-knowledge, rigorous self-cultivation, contemplation, and meditation. It can take many lifetimes. But some systems say, beautifully: Don't worry, there are many lifetimes to count. In the West, we're very nervous if we don't make it in this life. I think it's a limitation, that there is no possibility beyond this.

The whole question of ethical responsibility becomes critical here. In other words, as a cocreator in the process of understanding, we do not simply enhance our satisfaction in obtaining more knowledge for the human community. As actors we shape, and in a way control or dominate, the process of the cosmic transformation. I'm reminded of one passage from ancient Chinese tradition: If we can fully realize our own heart and mind, then we can understand our own nature. If we understand our own nature, then we understand the nature of human beings. If we understand the nature of human beings, then we can understand the nature of things. If we understand the nature of things, then we can take part in the transforming and nourishing processes of heaven and earth. If we can take part in the transforming and nourishing processes of heaven and earth, then we can form a trinity with heaven and earth. We can now read the whole thing totally in reverse. If we don't try to understand our own nature; if we insist upon invading nature for our own limited, egoistic understanding; if we develop instruments to harness the universe without deeper understanding of it; then in fact we become destroyers rather than cocreators.

We look at the great stars and the cosmic process and wonder what is the significance of a single individual, or even of all scientists, the entire human community. But if we take this particular point of view of our responsibility as cocreators, then human beings become a viable species on our own planet. What we do in the privacy of our home or in the confines of our lab is not only scientifically significant, socially significant, but also, in this very specific sense, cosmologically significant.

DAVID FINKELSTEIN: The conclusion to be drawn from the kind of theory that I've described, if it turns out to survive the week, is that far from being strangers in the universe, we are actually part of the law that governs it, and we help make the law that determines our own lives. Now, we mustn't exaggerate this. Our actions affect the space-time only slightly. Space-time is the stiffest medium we know. I'm sure that there are similar, very large numbers involved in estimating the effect of our actions on the laws of nature. We're presumably talking about very small effects, at least in the ordinary domains. But it is possible through this for us to feel at home in the universe. Andrei Sakharov, the Russian physicist who received the Nobel Prize, expressed some of the same ideas that Weiming presented in a remarkable equation. He wrote this as the dedication in a book that he gave to his wife: "The root of truth is love." It may not be generally recognized, but I know that if Sakharov wrote a square root, he meant a quantum square root. He is connecting two worlds: the old classical idea of objective truth, on one side, and on the other side, the world of action. He is realizing that truth is not, in fact, an objective thing. It's something you can only find if you love it very much because it takes a certain amount of effort.

The old idea of truth is that of a physicist outside of the world, like God, making theories about everything else. The actual situation that we see in the quantum laboratory is the physicist working like a dog to understand the least little bit of the universe. You saw how much equipment poor Anton had to bring from Innsbruck to show you one photon.

ANTON ZEILINGER: That was not much equipment. You have not yet seen what we call much equipment. [*This brought a smile to everyone's face.*]

DAVID FINKELSTEIN: If you want Anton to show you two photons, you must go to Innsbruck. He cannot bring them here.

What we look at is always an infinitesimal part of the universe. As the system grows, the experimenter must grow geometrically. Whenever we say anything, most of the world is implicit, unmentioned, perhaps even unrecognized, because we are focusing our attention on a very small region. Things like love and meaning are presumably not there under the microscope. But we shouldn't be surprised that we don't find them there because they are behind us in the home from which we came.

As David suggested, in June of the following year Anton and I joined the Dalai Lama, Alan Wallace, and Thubten Jinpa in Innsbruck, where we toured Anton's laboratories and then spent two glorious days on the mountainside overlooking Innsbruck, discussing the foundational questions concerning quantum mechanics.

David's beautiful closing remarks about the narrowness of our attention in physics can remind us of the larger issues that circle our scientific investigations and which were an integral part of our conversations with the Dalai Lama. Our scientific attentions focus on one small aspect of a vast universe in which we live out our lives. One can hold out the hope of enlarging our circle of relationships to include in our considerations the features not under the microscope: love and meaning. Earlier in this session, Weiming Tu invoked the image of a partnership between humanity and nature, or of a "joint venture with reality." What he and others seem to be pointing toward is the possibility of overcoming the delusion of consciousness that places us with our thoughts and feelings over and against, and forever separated from, the rest of the natural world. Compassion— "suffering with"—is such an overcoming of the divide. In a participatory view of the world, we become cocreators in some sense. The subtleties of quantum entanglement and observation begin to sound like part of a mature Buddhist philosophy.

I I

Science in Search of a Worldview

Piet Hut's research in astrophysics centers on the "million-body problem," encountered when galaxies collide. Using special-purpose high-speed computers that he and his collaborators in Japan and in the United States designed, Piet is able to model the evolution of galactic collisions with remarkable detail and accuracy. His opening remarks presented some of the ideas and spectacular images that are part of his cutting-edge research. But for many years, Piet's interests have reached beyond the technical issues of astrophysics to the philosophical issues arising from science itself. In addition to his research into the evolution of the cosmos, Piet has had a long-standing interest in what he terms "worldviews." This has drawn him into a deep study of continental philosophy (particularly the phenomenology of Edmund Husserl) and also of certain strands of Buddhist philosophy. Consistent with his interest in phenomenology, Piet will draw us back again and again toward experience in order to assess its proper role in science and in shaping our worldview.

Piet began by showing a photograph of colliding galaxies from the New York Times *of October 22, 1997.*

PIET HUT: I am delighted to be here today, and this whole week. I was very happy already on the flight from New York to Delhi because I saw this picture in the newspaper of the very things I study. It's a photograph, taken with a normal optical telescope, of two colliding galaxies. I am going to show you the same thing happening in my computer simula-

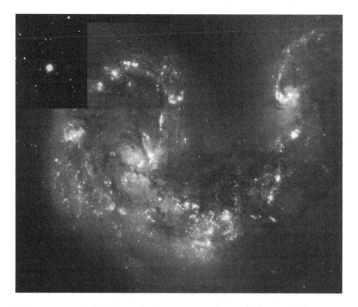

Figure 11.1 Colliding galaxies as seen through the Hubble space telescope. Courtesy of NASA and STSci.

tions of imagined galaxies, but it is nice to see the real thing here. You see a very complicated pattern of light, with two tails of light going out in different directions. We think that these are two galaxies that had a traffic accident. Because of the high speed of the collision, some of the stars were pulled off in these tails. We did not know that these were collisions until thirty years ago. Before that we thought they might be very strange galaxies, perhaps born that way. But thirty years ago, computers began to be fast enough to simulate the history of a collision.

What you see here [*figure 11.1*] is a picture of the same type of collision taken from the Hubble space telescope from outer space. Because there is no air, you can take much more beautiful images than you can from the ground. Not only do you see there is a collision, but you see gas clouds in the galaxies colliding and new stars being born because of the high pressure. Each of these many bright lights is a new star or a group of new stars.

DALAI LAMA: Do the stars actually collide with each other physically, or is it simply an interpenetration without actual collision?

PIET HUT: The stars don't hit each other. They are very small compared to the enormous space in between. So, two galaxies can pass through each other, although they still distort because they feel the gravity. But in between the stars there are big gas clouds, and these clouds collide.

DALAI LAMA: What happens to the direction of the motion? Is there any indication that one will pull the other in its direction?

PIET HUT: If they go at high speeds, they go through each other. They are a little bit slowed down but they keep moving.

DALAI LAMA: When these new stars are formed, do they ever form their own autonomous organization as a new little galaxy, or are they always dragged off by one or the other colliding galaxies?

PIET HUT: That's a very good question and both can happen. If you look carefully, you see that in these arms there are some bright spots where new stars are born. They will form small galaxies, child galaxies which will leave the main galaxy.

At high speed two galaxies pass through each other. At slow speeds they just stick. At intermediate speeds they don't pass through completely, and when they fall back they become one galaxy. That is what happens most often.

If you would like to see whether they go through or stick, you have to wait a few hundred million years. Most of us are not so patient, so we ask the computer what will happen. I will show you an example of a simulation.

DALAI LAMA: In that time, chances are that an individual could get enlightened and be able to tell the whole story of what happens to the galaxy.

PIET HUT: So, there is a competition between science and enlightenment.

Computer Simulations as Valid Inference

This last exchange brought laughter from us all. The other participants and observers then gathered around the computer screen, watching over the shoulders of Piet and His Holiness to see the simulation of colliding galaxies. The images we saw had been generated by one of the world's fastest special-purpose computers and were amazing to watch. As they unfolded, their beauty powerfully impressed us. At the end, spontaneous applause broke out.

PIET HUT: These are two galaxies at time zero. The time is measured in units of 100 million years. Both of these galaxies, the blue and white, indicate the stars. The red light indicates dark matter, the invisible mass that we know is there because we can feel the gravitational pull. If you look through a telescope, you see only the blue galaxies, like pancakes. Let's see what happens when they meet each other. So, you see the tails were similar to those in the photograph. The whole process, which you saw in a few seconds, took 400 million years.

DALAI LAMA: This is mathematically calculated? Is it more than sheer speculation?

PIET HUT: Oh yes. For each step of 3 million years, we calculate the gravitational force from each point to every other point. We have more than 10,000 points, so for each step we calculate 100 million forces. Actually, we cheat. At the start here, we add many forces together and compute them in one step instead of calculating them all separately. You do not have to get it absolutely correct. It is good enough to calculate the basic forces, but the mathematical equations are 100 percent accurate. The initial positions are the only question. We try many different positions until we get something that looks like what we see. It's like going into a forest for only one day to find out how trees grow. You can count the tree rings and think about it, but you only have one day to look at the forest. We only have one day in the cosmos, a very short time for analysis.

DALAI LAMA: I wonder if a simulation like this would be considered a genuine inference in terms of Buddhist epistemology. Genuine inference is real knowledge based on reasoning, where you can be very, very confident to an extremely high degree of probability that it is true. Is that the case, or is this an informed guess?

PIET HUT: I would say it is a good approximation. As an approximation it is very secure, but it is not secure for all the details. There are small aspects of gas clouds or magnetic fields or other things that we do not know.

DALAI LAMA: Are these events taking place within the region of the afterglow?

PIET HUT: The afterglow is radio radiation everywhere in the background, throughout the universe. It comes from far away for everybody. The afterglow is like the light of the sky, and these galaxies are like birds flying in front of the sky. The sky looks far away. Its light, if it travels to our eyes, is nearby, but it comes from far away. In the past, the whole universe was glowing, and after it stopped glowing the light kept traveling.

To show how the universe expands, we have to start with one little piece of it—several galaxy clusters—and watch it get larger. What you see is that the original gas in the universe forms small galaxies and gas clouds, and those galaxies and gas clouds fold together to form groups of galaxies. This is how structure in the universe is made. Each point of light that you see is not one star but many, many stars.

DALAI LAMA: Is there a typical number of galaxies within a standard galaxy cluster?

PIET HUT: They come in all sizes. Our local galaxy group is small. It has two big galaxies, ours and another one, and about ten or fifteen smaller galaxies. But nearby is a much bigger group, and we are part of the next hierarchy, a metagroup of thousands of galaxies.

DALAI LAMA: A poetic text by Nagarjuna asks, if the earth, mountains, ocean, and stars all eventually become mere dust, what cushion is there left for us weak mortals? What immortality can we expect? Even the galaxies dissolve. Astrophysicists should make this point especially to the politicians and to those people who are killing each other over religion . . . to what purpose?

PIET HUT: If we could get radio signals to other planets and astronomers could talk to each other as ambassadors, maybe we would have a better universe. But we have to wait a bit for that.

I very much enjoy playing with these simulations. As a child, I enjoyed playing with model railroad trains. When I grew up, my toys grew up, and now I play with galaxies.

ANTON ZEILINGER: What's the next step then?

From World Systems to Worldviews

PIET HUT: Maybe now I have to play with world systems, or perhaps worldviews. The nice thing is that I even get paid to play with these galaxies, but I also do some other things on the side. One of those is that I am very interested in the whole question of worldviews—views not only of the physical reality but of the world as a whole, including human beings and their sense of beauty and meaning.

In a dialogue between science and Buddhism, or more generally between science and religion, people talk about building a bridge. You mentioned recently that Nagarjuna says that many metaphors are only partly correct. I think that the bridge metaphor is partly correct, but I do not like it very much. I think the real meeting point happens when we go down into the canyon, to the roots where the knowledge of science and of Buddhism comes from, to this lesser known area. In America, where you need slogans because Americans have very short attention spans, I call this "roots, not fruits." That means a focus on the process, not the results or the fruits of science and Buddhism. It is very interesting to talk about the big bang or about Buddhist and Christian equivalents of the origin of the world. But even more interesting is the

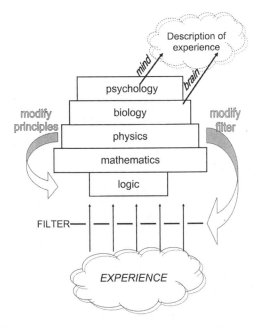

Figure 11.2 Experience and the hierarchy of knowledge

process of knowledge formation, of the development of wisdom and of compassion. Maybe in science in the future we'll be able to talk about that process of knowledge.

PIET HUT: If we want to go to the roots of science and religion, as a scientist, of course, I have to start on the side of science. I would like to show you my view of the roots of science [*figure 11.2*]. I first have to look at the structure of different forms of science, at their relationships, and the process of how they grew. Sometimes people use the metaphor of the house of science, with mathematics on the first floor and physics, which is based on mathematics, above it on the second floor. Biology uses physics and chemistry, and these are like the different floors in a house. Mathematics itself is based on a particular logic. I have made the foundation of logic smaller because there is a limited number of logical rules from which a large body of mathematics and physics extends. As a result it looks more like a *stupa* than a house, but that's okay in this environment.

There is a reductionist tendency in this model—a tendency to explain life by looking at DNA, at the physics or chemistry of molecules and atoms. Similarly, if people study the brain in psychology, they look at the biological structure of the brain, at the nerve cells and how nerve signals propagate. Each level is explained in terms of a lower level, from psychology down to biology, down to physics, down to mathematics, down to logic. Scientists, as well as people outside science, often talk about science by starting with principles. There are scientific principles, such as objectivity, repeatability, and consensus between scientists, which then lead to logical, mathematical laws. Everything else rests on top of that. The outcome of biology and psychology is, for example, knowledge of the body and the brain and more and more knowledge about experience—how human experience arises from the brain, from the embodied being, starting with the simple elements.

This seems to be the structure and the process of science, but I think this picture is misleading. It is too simple. It is an ideal, but in practice what happens is that new discoveries sometimes force us to modify the principles. In physics, for example, it is true that at any given time logic determines how to do the mathematics, and mathematics determines the physics; there are these relationships between the levels. But new discoveries like quantum mechanics force us to change the first principles, to start with new principles to explain what is happening.

DALAI LAMA: Historically in Buddhism, it is as if physics were to come first and a formulation of logic were to come later. Buddhist logic was strictly formulated in the fifth century by Dignaga and then reformulated in the seventh century by Dharmakirti, and Buddhism was long established before that. It's interesting to consider what the basis of logic is. The law of the excluded middle, for example, is based on a kind of experience. As we look around and see how things happen, on the basis of that experience we start to formulate logic. The logic wasn't made in a vacuum. It was an embodied logic to start with, which we draw out from experience. Then the logic comes around and defines and clarifies further experience.

PIET HUT: It's very similar in physics now.

At this point, Thubten Jinpa and the Dalai Lama became involved in an animated conversation in Tibetan, which Jinpa then explained.

THUBTEN JINPA: I was arguing that logical principles, such as identity, contradiction, or the law of the excluded middle, have a lot to do with the way human thought functions. They are based simply on the tools with which thought operates, and they are fundamental assumptions without which we just can't make sense. His Holiness took the opposite position, saying that these are principles abstracted from the physical world. For example, the quantum phenomenon of superposition suggests that there may be a need to modify the Buddhist logical position that the law of the excluded middle applies in all instances.

DALAI LAMA: If logic really isn't coming simply from an a priori position, disengaged from nature, but is based upon nature, then you have to modify your logic as new information comes in. Buddhism must modify its logical principles based upon the new empirical evidence coming in.

PIET HUT: I think physics has to change much more than Buddhism. David has expressed some very interesting ideas about the need to change to a quantum logic. It is a very exciting topic.

What you just said about the historical origins of Buddhist logic reminds me that this house of science began to be built very recently, only 400 years ago in Europe. They had inherited the logic and mathematics of the Greeks. That was already at least 2,000 years old. But the Greeks were only talking about static things. Then Galileo and Kepler discovered the physics of motion. It was clear that to understand this new physics, such as the detailed motion of the planets, we needed to modify the mathematics and the logic used for their description. Newton invented a new mathematics for very small intervals—in principle, for infinitely small intervals—to be able to describe arbitrary shapes. Newton investigated the first modification of mathematics based on principles from physics. People thought that after Newton there was no further need for change. This was dogma until quantum physics.

Of course, this classical framework is still correct to a high degree, but if you do very precise experiments, you have to modify it. So, it is still very useful, but not entirely accurate. In quantum mechanics, for example, you cannot exactly repeat something, such as the time taken by a radioactive atom to decay. Three hundred years ago, the most important thing in physics was repeatability. If you could not repeat something, you could not verify it. That really was dogma, a cornerstone of science. The strange thing about the building of science is that it is like a floating house: You can change the ground floor or the basement while the rest of the structure still stands. It's not entirely true that psychology has to be based on biology, and biology has to be based on physics; if it were true, then modifying the principles of the foundation would cause everything to collapse. But that does not happen. Biology has its own understanding, and even if you change physics, it will continue. There is no reason to believe that this process of feedback, leading to modification of principles, will not continue. Just as physics has led to new mathematics, I expect that biology or psychology will lead to new physics and make the foundation richer.

DALAI LAMA: Are you talking here about empirical psychology?

PIET HUT: Yes. Empirical first and then theoretical deductions follow from the empirical results. Theory and experiment always go together. Theory alone is dogma. Experiment alone is mute. You do not even know how to describe an experiment if you don't have the language of theory to talk about it.

Other examples of modifying principles show that the general tendency is not arbitrary. There is consistent pattern. It always moves in the same direction, namely, from absolute to relative. If we learn something new, we first think, "Ah! I have found the truth." Then we learn more, and we think, "Mmm . . . maybe it is not absolute. Maybe there is another side to it." We don't throw it away, but we place it in a larger context. In physics, we have even called such developments relativity theory, as in Einstein's relativity theory. But calling a theory a relativity theory is a negative way of saying it. A more positive way would be to call it a transformation theory.

Transformation Theory

PIET HUT: One of the simplest relativity theories is kitchen relativity, where you can transform water into ice. You can freeze water or melt ice. If you grow up on a tropical island, you don't know ice; you only know water. Water is absolute; you never see anything else. But one day you travel or you get a refrigerator, and you can transform water into ice. Then you see that water is relative. Ice also is relative, but the material seems to be absolute since you can transform in either direction. It is the same material but with a different appearance, a different form. The relativity of water and ice gives you more freedom. It gives you the freedom to make a transformation. You can do more things than you could before. Similarly, Einstein's relativity showed us that space and time are not absolute, but you can transform them to some extent into each other.

DALAI LAMA: In the case of ice and water, you simply had different manifestations of form and the same substance. Is it really so closely analogous when you speak of time and space? Does time actually transform into space, and space actually transform into time?

PIET HUT: It is a close analogy in the sense that, normally, what is space from one point of view is space with a little bit of time taken into account from another point of view. Likewise, what is time for me will be mostly time with a little bit of space mixed in for you. This only becomes significant at very high speeds.

DALAI LAMA: It looks like simply two corollary changes taking place—this changes; therefore that changes—rather than an actual transmutation of a body of space into a unit of time. However, the substance of the water actually turns into the substance of the ice and vice versa.

PIET HUT: Actually you can make a complete transmutation from space to time. But you can only do that inside a black hole.

DAVID FINKELSTEIN: There is no symmetry at all between time and space. You were correct when you said that space enters into time but doesn't become time. Time enters into space but doesn't become space. It's as if you could transform a pitcher of water to a pitcher of water with a little bit of ice in it. But you can't go all the way.

PIET HUT: I would argue that time and space reverse roles with respect to a distant observer. But let's talk about that offline; it is a rather technical point.

DAVID FINKELSTEIN: Going back to the example of the railroad trip, if I make a round trip I have moved in space and time. From another person's point of view, I moved only in time. We can have a transformation of time into time and a little space, but never time into space.

PIET HUT: What can transform completely are mass and energy. You can have mass completely transform into energy and energy completely transform into matter. A nuclear bomb and a nuclear reactor are examples. That follows from the space-time transformation. If you study the mathematical theory of Einstein's relativity, you see that space and time transform at least to some degree and that mass and energy transform fully, like water and ice.

DALAI LAMA: We have a lot of illustrations, even in the macroworld, of mass turning into energy. It happens in the fireplace. Can you give an example of energy transforming into mass?

PIET HUT: You can see it happening in elementary particle processes. A photon is transformed into two material particles: an electron and an antielectron. Material is produced from pure energy, from a photon.

DALAI LAMA: Does this happen only at the microscopic level?

PIET HUT: It happened with the whole world after the big bang. After the big bang, everything was energy in the form of radiation. When it got colder, the radiation condensed into matter, like steam condensing into droplets of water.

DALAI LAMA: But it does not happen as a day-to-day event in the macroworld?

DAVID FINKELSTEIN: When a plant absorbs solar energy, it gets a little bit heavier.

PIET HUT: Yes, in principle, but by a very tiny amount.

DALAI LAMA: How can you have energy all by itself without some material basis for that energy? Likewise, can you clarify how a photon has no rest mass, but when it's in motion, it does have mass?

PIET HUT: The problem is that you could say the photon is a type of physical material. So the modern view sees a photon as a form of matter. A hundred years ago, people said that light is pure energy, and this table is pure matter [*knocks on table*]. Before relativity theory, there was a duality between matter and energy. But now we know they are similar things.

DALAI LAMA: How can you have energy just standing all by itself, without a material source?

DAVID FINKELSTEIN: There's no such thing. Energy is always a property of something else.

DALAI LAMA: If there is no such thing as energy standing all by itself, what does it mean to say that energy transforms into mass?

DAVID FINKELSTEIN: I think we have to say it correctly. Your Holiness is finding logical errors in our ordinary way of speaking. It's always simply a transformation of one form of energy into another. It's just a very old-fashioned way of thinking that mass is turning into energy. Turning mass into energy is like turning pounds into ounces. It's the same stuff. There's energy in this cup, but it's bound into the nucleus. They are exactly the same thing; they differ only in units.

ANTON ZEILINGER: I would disagree with that because I have to define things operationally. What do I mean by mass?

DAVID FINKELSTEIN: You can measure both mass and energy with exactly the same instruments. That's Einstein's discovery. You can weigh energy, and you can measure mass with a kind of thermometer, a calorimeter. They are the same stuff. It's not even a transformation—just a liberation, a change in form.

ANTON ZEILINGER: You're right. You can weigh energy.

DAVID FINKELSTEIN: Matter is not the same as mass. Radiation is not the same as energy. Radiation is a cloud of light particles. It has energy, just like it has color. Energy is one property of it.

PIET HUT: The confusion is because we started with one understanding 100 years ago, and we continue to use the words derived from that understanding. Now we have a new understanding, but we still use the old words.

The reason for the confusion about the rest mass and moving mass of a photon is this: If an object moves, some of its energy is the energy of motion. Much of its energy is mass, locked up in the material, which could, in principle, be released in a nuclear explosion. If I want to find out how much mass, it's hard to know if the object is in motion. I have to stop it and carefully weigh or analyze it. The problem is, if you stop light, it disappears. You cannot stop light. In a photon, all of the energy is in the motion. There is no energy locked up. A photon is manifest energy. A physical object has hidden energy.

The questions posed by the Dalai Lama about the nature of matter and energy carefully probed the ontological commitments of the scientists in the group. Not surprisingly, each of us has a somewhat different view of the nature of matter and energy, mass and radiation. The physics of these aspects of our world is largely understood, but the philosophical implications of our most advanced understandings of such areas as relativity and quantum field theory are still being debated. The Dalai Lama was immediately interested in exactly those issues that are still under discussion.

Following this exchange Piet returned to his presentation on the structure of scientific knowledge and the role of experience.

The Role of Experience in Science

PIET HUT: In our picture of the house of science, we saw that scientists normally cheat and tell you only part of the story. They do not often tell you that they rebuilt the foundation. Another thing you do not hear about very often is even more fundamental: the filter that separates experience from the construction of science. For me this is extremely important, and it is almost always left out of academic discussions.

DALAI LAMA: By filter, do you mean exactly which aspects of general experience we will filter out and which we include as part of science?

PIET HUT: Yes. The primary and secondary qualities are an example. Three hundred years ago, people determined that the length of an object is physics, but its touch and color is subjective. Human beings can feel the object and can see the color. But in physics, we only talk about mass, length, and time. Color has not been interesting for physicists. Now we have a much more detailed understanding of matter, and we have modified the filter: Now we can compute the color of materials. Our filter is getting larger, and we can describe more.

DALAI LAMA: But even now in physics, when you speak of color, you are talking about photons and such things. As Arthur pointed out with his study of Goethe's color theory, you're still leaving out what we actually experience as color or sound and so forth.

PIET HUT: Subjective experience does not go through the filter. Beauty and responsibility and meaning do not go through, at least not at the moment.

DALAI LAMA: Does mind go through?

PIET HUT: Not as subjective experience. When scientists talk about experience from the standpoint of psychology and biology, they are focusing on the body and the brain. While this "experience" comes from the real experience, it leaves out much. Then they make an abstract picture, using mathematics and physics. They build up to biology, and then they reconstruct the experience. There is no reason to believe that it works completely. It is only an approximation.

If a neuroscientist tells you that he or she knows this or that about experience, or if a biologist claims knowledge about human brains from evolution, specific conclusions may be right. We have a lot of detailed knowledge. But there is no reason to believe that we have the complete picture. Probably we do not because so much is left out and the knowledge structure is constantly changing. But as the filter is being modified, then hopefully our understanding of experience is improving and getting more accurate.

What I think is most interesting about science is the notion of freedom from identification. In this century, we have seen that the old picture of the world of objects that we see around us really has to be replaced by an interplay of interactions. Every phenomenon is an interaction. Everything we know about the photon is given as a play of actions. The photon can sometimes play as if it is more like a wave, and sometimes it plays more like a particle, depending on which question we ask. We cannot identify it uniquely, saying an electron is a wave or is a particle. It is more fluid; there are more possibilities. Using our understanding of different roles, we have to see that the roles are only roles and not definitive, not absolute. Therefore, at least in physics, we can see that we really need to give up identification.

DALAI LAMA: By identification do you mean the notion of thingness?

PIET HUT: Yes. Identification of any fixed thing is one example, but also any property. It's like the elephant. You cannot identify it with a leg or with a tail or with any specific set of parts. The change from objects to actions, or from objects to phenomena, can be considered as a change from "is" to "as." An electron *is* not a piece of absolute substance. But an electron can appear *as* a particle or *as* a wave. It can play a certain role. One way of saying it very briefly is to say "nothing yet there" or "nonexisting it appears." I really like that expression. I think that this truth was discovered on a very fundamental level much earlier in your tradition than in the scientific tradition. I do not think it is a coincidence. I think it tells us something about the fundamental structure of reality on the level of the roots, not the fruits.

DALAI LAMA: It can't be a pure coincidence that both science and Buddhism come to more or less the same conclusion on the nature of emptiness when taking physical objects as the focus of analysis.

PIET HUT: In principle it could be a coincidence, but only if the physical world and the mental world are absolutely different, without any possibility of transformation. But I think that we need to look at possible connections between relativity and transformation. And if they are connected, then I do not think it is a coincidence.

As an example, we talked yesterday about the big bang in physics and space particles in the Kalachakra system of Buddhism, and it is very interesting to compare those two. Some aspects will be similar and some may be different because the results are determined in part by the method of investigation. However, I think the logic is most fundamental. If you really understand the logic more and more, at some level the ground has to be similar. If it is not the same, we have to go deeper to find the connection. Somewhere the canyon must have a bottom that connects both sides.

Already I think science has found a high degree of freedom from identification. The question is, Where are we going in future? I can only see that we are moving from a science of objectivity to a science that includes subjectivity, as well as objectivity. The next relativity theory or the next transformation theory will include a relativity between the object and the subject, between the physical and the mental.

DALAI LAMA: It's getting more encompassing and vaster. For a long time my intuition has been that up until now the domain of science has been rather confined in being limited to the physical world, focusing only on what can be quantified. Gradually science will have to expand its horizons so that it can bring into its domain of analysis phenomena that may not be subject to quantification like physical objects. From what you have shown us, I feel you may share the same hope.

PIET HUT: I have the feeling I am climbing down from science's side into the canyon, and the deeper I go the more I can see the other side. I cannot jump yet. I am a little bit too scared to make such a big jump, but from here I can see the Tibetan notion of the sameness of outer and inner space—that they are not really something different. I recognize the language from the other side, and I see in it something very similar to what I expect to happen in the language of science in the next hundred years or so. The search for a wider view, a wider context, a wider space —that is what science will soon investigate in much greater depth. It would be very nice to look at Buddhism and see whether we can get some help. In the beginning it was very difficult to help each other. We were too far away, at the top of the canyon, but now we are getting closer. It is more and more possible to learn from each other at the level of the underlying logic and processes.

In Piet's discussion of experience, he, like others in our group, articulated a view in which the objectification of the world is moderated to include subjectivity in an appropriate manner. In place of the radical schism between inner and outer, between subject and object, Piet proposed that we explore views in which lived experience is granted a more important place in science than previously. Perhaps we will even come to a new relativity or transformation theory that will show us the way to move correctly from one to the other, from the objective to the subjective, for example. In an important remark, the Dalai Lama agreed with Piet. He indicated his own intuition that the science of the future will have to be more encompassing, including within it phenomena that are not purely material and which cannot be quantified in the same manner as physical objects.

This view is very much like my own. I am convinced that we can have concrete and detailed knowledge of a much broader array of phenomena than a strictly quantitative and materialistic view of science permits. The word science *stems from the Latin* scientia, *meaning "having knowledge." Like Piet and the Dalai Lama, I believe that it will become increasingly important to recognize that we can have knowledge of a much broader range of phenomena than science has traditionally allowed. This includes knowledge based on lived human experience, both of the outer world, accessible to the senses, and the inner world, opened by reflection and contemplation. In other words, the scope of science can indeed become more encompassing and vaster in a way that is neither reductionistic nor strictly quantitative, and yet it can remain true to the essential values of scientific inquiry.*

Between Illusion and Reality

ARTHUR ZAJONC: Piet, could you amplify one of the points that you have raised about the play of actions? You talked about how, through increasing relativity, we free ourselves from the tendency to identification and reification of the world of objects, focusing instead on a play of actions. One of the dangers here is that the language you used suggests whimsy, that it is a mere play of appearance or actions, as opposed to something that has content or meaning. There is an analogy in relativity theory in that space and time considered independently are illusory. Here, also it is not the case that they are just playful appearances. There still is a deep structure, but the structure is at another level. Could you say something about this middle ground between objectification and complete relativism?

Although I addressed the question to Piet, His Holiness stepped in.

DALAI LAMA: I want you to continue, but let me add something. The Madhyamika Prasangika view posits a form of knowledge that can also have an aspect of illusion in it. One can establish valid cognition even though appearances have an illusory quality or aspect to them.

Within the Buddhist epistemological discourse, there are two divergent opinions. Some maintain that any form of valid knowledge must be valid in relation to all aspects of the object, which implies a belief in some kind of intrinsic being of the object of perception. This is a classic correspondence theory, with a real world out there corresponding to valid cognition. However, the Madhyamika Prasangika view argues that there is no need to attribute a true being to the object, and also there is no need to assert that knowledge is valid in relation to such a true being. One can talk about an illusory knowledge. From the other point of view, the discussion of interplay or appearances immediately suggests hallucinations and pure illusion. But once we are able to accept the notion of valid cognition that may have an illusory appearance, then there is no danger of misinterpretation when we talk about appearances and mere play.

PIET HUT: You gave a very nice example a few days ago, about looking at a flower — that if you believed the flower was really there, you could appreciate it in one way, as substantial. But if you realize the flower is not substantial, then you are even more free to appreciate the flowerness of the flower without having this fixed identification. I think there is a similar aspect here of freedom from identification — not having to grasp but just appreciating the phenomenon and the structure in the phenomenon.

If physics changes the rules of the game, it can change the principles and the filter, but physics does not have the power to change the experience before subject-object analysis. Insofar as you want to talk about reality, that nondual experience is most real, most given. There was a mathematician in America who asked his students if they really believed that by changing set theory, for example, they would cause the bridges built on the basis of that theory to fall.

DALAI LAMA: Whatever transmutations physics and the other sciences may go through, however they may redefine themselves, reality stays as it is.

ARTHUR ZAJONC: Is experience dependent upon the observer? Is it necessary to have an experiencer in order to have experience?

PIET HUT: I would say that, within the field of experience, there is the appearance of an observer and the appearance of the observed. The observer who has experience and the experience that has an observer are interdependent.

DALAI LAMA: There are two terms for experience in Tibetan. Perceptual or direct experience *[lengay]* means the natural, spontaneous, raw perceptual experience that you are born with. *Kundop* means something conceptually structured or fabricated. That doesn't mean it's fallacious, but it is more conceptual, derived from a great deal of highly theoretical investigation. This latter emerges as a domain of experience that would not have occurred without all of that research, investigation, and conceptualization.

PIET HUT: Looking from science, I would add that there are filters. In addition to the filter that science places on experience, there are other filters in our educational system, in our childhood, in our culture, in the way our innate experience is molded into a shape. For example, we believe in subject-object organization. The subject-object split is deep down inside us before it appears in science. My answer to Arthur's question is that deep down, it is nonduality. The subject and the observer are part of the totality of experience.

DALAI LAMA: Are you suggesting that starting from raw cognition, valid and invalid, you could go through training and research, until finally the process of science brings you to valid cognition? That is, in fact, what is sought in Buddhist meditative training. You start out with a lot of confusion and false assumptions, but you go through a discipline, and as a result of that discipline, you end up with valid cognition. Are you suggesting something comparable here?

PIET HUT: From the point of view of science, this is very confused. Science is making progress, but it cannot say anything yet about the original raw experience. Science can look at what Buddhism has to say about this experience and can get inspiration from Buddhism, but science can also continue on its own, modifying the principles and enlarging the filter. The wider scientific framework will look very different than the present one. Present physics looks very different from 100 years ago, and 100 years later, I would guess, it will look very different again.

In this exchange, both Piet and the Dalai Lama inquired about the forms and transformations of human experience. Piet had emphasized the modifications of experience that take place through the filters imposed by science but also, even more fundamentally, by culture. Piet maintained that at its root, experience is nondual, that the apparent dualism of subject and object is itself due to a filter or modification imposed on experience. Dualism is derivative, not essential. In response, the Dalai Lama quite naturally underscored the Buddhist view that we can change the factors that affect our experience, that we can modify the filters, for example, through disciplined contemplative practice. Thus, through an appropriate schooling, one can move from raw cognition to increasingly refined and disciplined forms of cognition until one ends up with a valid form of direct cognitive experience. The confluence of opinion expressed here resonates with other parts of our conversation, especially the comments made by Tu Weiming. In the following final session, several strands of our week-long conversation are pulled together.

I2

Knowing and Suffering

Clothed in facts
truth feels oppressed;
in the garb of poetry
it moves easy and free.
—Rabindranath Tagore

We opened the concluding session with remarks from one of the guests in the audience, Eiko Ikegami. Eiko was then a professor of sociology at Yale University and is now at the New School of Social Research in New York City. Her remarks represent the sentiment of others attending the meeting, who were interested in the relationship between religion and academic research. But Eiko also spoke with the voice of a scholar who knew firsthand the American, as well as the Asian, point of view.

EIKO IKEGAMI: As a Japanese, as well as a social scientist, what has been most striking to me in this five-day conference is an attitude that focuses on inquiry rather than just acquiring knowledge. I was so impressed, as all the guests have been. The scientists here are tough students, but debate and critical discourse are also part of the tradition of Buddhism— a very valuable part that shows us that appreciating experiential knowledge does not mean we have to drop rational discourse. That has been most valuable to learn, not only because it furthers the dialogue between religion and science, between East and West, but it also furthers the social dimension of knowledge. Ninety percent of my colleagues in social science, political science, sociology, economics, and so forth are still under the influence of the classical nineteenth-century scientific view, where the quantitative aspect of analysis is most important. For them Asian thought is an object of research, not an opportunity for learning, unfortunately.

Most of them, unfortunately, are not very different from those in the physical sciences. Actually, the same is true for many Asians. Since the nineteenth century, we have been looking to the West as a source of progress because we have to modernize our society. Japan happened to be historically the earliest and fastest to learn Western science and technologies. We in Japan also face the negative side of that quick modernization. Most Asians value tradition, but traditions themselves sometimes have a negative impact on social dimensions. This is unfortunate because we value the modern age, critical discourse, and the rational scientific mode of inquiry, not only as a source of technological progress, but also as a foundation of democracy. It would be very difficult now to drop the critical, empirical mode of inquiry in the modern life in Asia. It would be impossible. Both social scientists and Asians have a common fear that, if we had to drop critical discourse and empirical inquiry in order to accept or appreciate the value of meditative or spiritual experience, we could not do it. But over the last five days, we have observed that if we go to their roots, both science and religion are approaching a similar goal, and critical discourse remains valuable as a mode of inquiry.

DALAI LAMA: Your words are a source of encouragement to me. Even among the Tibetans, we find some scholars who feel that science and religion are so far apart that one has hardly anything to teach the other. They simply dismiss science as irrelevant, just the opposite from some people in the West who dismiss religion. I would like to replace the word *religion* with *religiosity* because it is not the specific discipline that matters here but an experience which is more general than any one religion.

PIET HUT: I appreciate what you are saying about religion and religiosity. But even religion that includes fixed, dogmatic structures has something living, something experiential, if you go to the roots. It always starts from a living ground.

In science, if we ask for an integrated worldview, we really find nothing. Human values do not have a place in science, at least not yet. Beauty or meaning does not exist for science. Strictly speaking, science does not have a worldview, only a small, partial view. But science is universal. It can be shared across many countries, many cultures, and it has the possibility for growth. For 400 years, science has been accumulating, growing, and modifying its own principles by its own power. Religion has an integrated worldview. There are different types of religion, of course, but all of them strive for a complete view of human beings, as well as the world. But the questions of universality and growth are more problematic for religion, or at least more complex.

I hope we can explore how to take the strengths of science—its universality and growth—and the integrated worldview that is a strength of religion and spread each to the other. I so often see people limiting their view. For example, many of my colleagues think that science is a worldview and that a reductionistic description of human life is enough. Other people who are religious, for example, many Christians in Europe and America, look to religion for a worldview, even as they look to science for universal and ever-growing knowledge. I hope it is possible, in a dialogue between science and religion, to find a universal element of religion, at least at its roots, regardless of the many forms that religion can take. And I hope that the growth of knowledge can be found in this universal domain. I have seen periods of growth in many different religions, certainly in Buddhism. But there are also periods of stagnation, I think, in many religions. So I hope this exploration is possible.

DALAI LAMA: For all of this, it's imperative to engage with the material with an unbiased, open, unprejudiced mind. In the Buddhists treatises, three characteristics define a qualified student, one who is called a suitable vessel for learning, for engaging in spiritual practice, for receiving teachings. One of those three characteristics is having an open mind, a lack of prejudice or bias. The second is being perceptive and intelligent, and the third is having a genuine aspiration or yearning.

These attributes of a qualified student would be excellent criteria for students in the West as well. As should be clear by now, the Dalai Lama himself, as well as all the scientists present, exemplified these qualities. Whereas each was competent in his or her own field of inquiry, each retained an open mind and was eager to learn from the others. This was true even if the modes of expression and the textual references spanned many fields, cultures, and centuries. The importance of remaining open not only to new facts or arguments but also to other ways of knowing was brought home by a lovely story, told by Tu Weiming, about the tension between logical argumentation and direct experience of a certain kind.

TU WEIMING: The central issue of fruitful interaction between analytical, critical thinking, on the one hand, which is embodied in Western science and which is now problematic in some sense, and the religious consciousness rooted in the East brings to mind an anecdote. A Taoist master and his friend, a famous logician, were standing on the bridge that crossed the river Hao. The Taoist looked down and said, "How happy the fish are!" The logician said, "How do you know whether the fish are happy or not?" The Taoist said, "How do you know that I don't know whether the fish are happy or not?" The logician said, "Because I'm not you, I don't know whether you know that the fish are happy. And you are not a fish, so how do you know whether the fish are happy?" The Taoist said, "I know the fish are happy by standing here now."

It seems that the logician, simply through inference, could not comprehend the aesthetic, or even mystic, experience of the Taoist in appreciating the happiness of the fish. The centrality of experience becomes critical here. Piet made the statement that the real basis for science lies in experience, before the various filters of science are applied. These filters determine what of the full human experience may pass through to the description that comes out of scientific examination and observation. Piet also pointed out the need to search for a wider vision. And the wider vision, in a way, is what Eiko later described as the fruitful interaction between the various filters, very rigorously designed in the analytical method. But it is also the ability to go beyond all these conventions and be truthful to what one actually observed, to open one's mind to all kinds of creative potential. It is in this sense that I think the observer, the knower—whom we now also call the experimenter, the actor, or the cocreator—becomes critical.

How do we think, and how do we know? It seems that scientists and others involved in the modern world can learn not only from religion but also from aesthetics and from many other domains of human experience. There is a challenge that is so fundamental: that we have to learn to know and to think not only with our head but also with our heart, and even with our body. That's part of the reason that some modern scholars have put a great deal of emphasis on personal knowledge, or what we could call embodied thinking. I must add that feminists have contributed greatly to this kind of inquiry because we need to develop an integrated, connected dynamic, as well as an extremely open and holistic process of thinking and knowing.

There is a single word in Chinese, *ti,* that means both "the body" and "to embody." This word very often forms compounds with other very common words and in doing so changes the meaning. Such common words are "to recognize," "to examine," "to probe," "to comprehend," "to verify," or even just "to taste." But if you combine these terms with the word *ti,* then you have "embodied recognition," "embodied examination," "embodied probing," "embodied comprehension," "embodied verification," or even "embodied taste." But if we change the word "embody," which is of course a bit awkward, into an ordinary English word, we normally use the term *experiential*. So it would be an experiential verification or experiential understanding.

What is experiential understanding? The Chinese have developed epistemological concepts about this, but the most fruitful area is in art. A very simple example from painting is the argument against the idea that a painter simply tries to depict what is out there. The most a painter is able to do by depicting what is out there is what the Chinese call form-likeness. Form-likeness describes not just the veracity but the childishness of the painter. A mature painter would be able to do something more than simply describe what is out there. His vision has to be more than simply photographic. The painter is enjoined, not simply to depict what is out there, but to be involved in a spiritual communion with the object of study, of understanding. It's not just a mystic notion. The familiarity with the object of study becomes so integrated into your own experience that it becomes an extension of your body. A master was asked by a student how to paint a mountain. The student said he had already refined all the techniques and knew how to do it very, very well. But the master was not happy with the product because it was just the form-likeness of the mountain. So the master said, "Forget about painting. Go to the mountains and live there for a while. Enjoy walking there to familiarize yourself with the mountain, to involve yourself in a spiritual communion with the mountain." The poets describe the mountain, and it's very much like the mountain here in Dharamsala. You never see the mountain, but you sense that you are in the midst of it. You do not see it because you are an inside participant. So, the idea is to ask the artist to cultivate a sense of taste, a feeling about the mountain. Then, when the artist is ready, he will be able to express himself in a kind of creative joint enterprise between what the mountain evokes in him and how he is able to respond.

In the early 1960s, the official journal of the Academy of Arts and Sciences published a special issue called "Toward the Year 2000." In it they made predictions about the decades still left until the year 2000. Some of the original authors are still around, and they're very happy with what they managed to come up with, but they realized they missed two things. One is ecology. Our vision of the world was very different then, before man landed on the moon and could look back at Earth. The notion of nature as out there, with ourselves as observers, still prevailed, and there was no real, not to mention deep ecological thinking. Deep ecological thinking is rooted in an experiential understanding of nature. The other point they missed, to their surprise, was the feminist movement and the changing role of women in the last forty years. If we look at it philosophically, that movement contributed a different mode of thinking based upon relationality—thinking that is not differentiated from feeling or sympathy; thinking contextually; thinking as an affective, as well as cognitive, mode. That mode of thinking suggests the possible global significance of local knowledge through intersubjective communication. It suggests what we see as a new horizon, or as Piet described it, the interplay between subjectivity and objectivity.

That interplay is linked to a notion of deepening subjectivity: the ability to understand the local in such a way that it has become almost like a well, to use a Sufi image. If you dig a well deep enough you don't bury yourself in the hole. You reach the common spring of communication. You do not begin the communication simply by lifting yourself up from the local position because you cannot do that. It's an abstraction. You very carefully give a thick description of the local knowledge that you have, all the positive and negative areas. But if you dig deep enough, into deepening subjectivity, you will reach a common spring. Your digging and David's digging and Anton's diggings and all the others' will meet somewhere. That communication is intersubjectively confirmed; it's rich and it's contextualized. It's not generalizable in terms of universality, but it is generalizable in terms of communication, in terms of dialogue and mutual understanding.

Tu Weiming managed to capture the hope of several of us, namely, that we would come to see experiential knowledge as an essential part of the larger, more encompassing project of knowledge spoken of by the Dalai Lama, instead of viewing it with open suspicion. This is not to abandon the rigors of science or to reject critical thinking and analysis. Rather, it is to enlarge the territory of knowledge and to diversify its character. The logician and Taoist philosopher both live in our breasts. To make peace with ourselves and our world, we need both—we need the linear and the complex, the logician and the ecologist, the masculine and the feminine. Nothing less will suffice.

Suffering and the Role of Knowledge

[*One of the guest observers asked*] Your Holiness, during the lifetime of the Buddha, he was often asked cosmological questions, and in most instances he dismissed those questions, without answering, as not relevant to relieving *dukka,* or afflictive emotion. Why do you feel now that these questions are so important to address and answer in the modern context?

DALAI LAMA: There are two answers to this. First of all, not all of the participants here are Buddhists, so we do not ask them to follow Buddhist principles or look to the Buddhist scriptures to see what should be pursued and what should not. The second issue is a larger one. Before you establish the path and set out for enlightenment, you need to establish what the nature of reality is. What's really going on? Cosmology is part of what's going on. The finest knowledge available should be integrated into the Buddhist picture of what's going on, and from that you can develop the path and proceed to the fruition. However, when I talk to the Tibetan contemplatives living up in the mountains in retreat, I'm not going to mention quantum mechanics. It's not pertinent to what they are doing. They don't need to know.

But historically, one must not forget that there are different interpretations of the significance of the Buddha's not answering these questions. One is the view you presented: that they have no relevance to the individual seeker on the path. Nagarjuna gives a different interpretation of the fourteen unanswered questions. He says that it has a lot to do with the context in which the questions were raised and the motivation and metaphysical assumptions in the mind of the interlocutor who asked the question. For example, one of the questions was whether or not there is a soul after death. If the person asking operates from a premise that there is such an intrinsically real entity, a positive answer from the Buddha would affirm his reification. If the Buddha were to answer in the negative, he would fall into nihilism, denying his own being. So, the refusal to answer these questions depends on the form in which they were raised and the assumptions behind them.

ARTHUR ZAJONC: Often it is the case that a powerful mode of investigation can become a source of suffering if we fail to fully consider its character and limitations. Scientific inquiry, which is in itself a very noble thing, can be distorted and become a source of pain and suffering in this way. We have been trying to understand science in such a way that it does not become a source of suffering. For by ignoring it, by not turning our attention to it, we allow only those forces and those intellects to dominate science which may end up causing suffering. By engaging in dialogue like this, by coming to wisdom rather than ideology with regard to science, we have the opportunity to achieve the positive rather than negative potential of these investigations.

DALAI LAMA: Once the student asked the teacher a question. The teacher did not know the answer to this question, and he responded back by saying that not knowing is a source of suffering. This suggests that we have to know.

Atisha, one of the Indian masters who came to Tibet around the eleventh century, made a hobby of doing crafts and making mechanical things. One day he was mending a broken clay pot. One of his Tibetan attendants passed by and asked why the great lama was doing this manual work. Atisha responded by saying, "Aren't we all supposed to be searching for omniscience? This is a part of knowledge."

We are our own master. Things depend entirely on us; they rest on our shoulders. Therefore, the future of humanity is in the hands of humanity itself. We have the responsibility to create a better world, a happier world, and a more peaceful world. I feel it is our responsibility.

I want to express my deep appreciation to you all. Thank you.

ARTHUR ZAJONC: I know I speak for everyone here in expressing our thanks to you also for hosting us and for engaging with us in these important dialogues concerning the deepest issues confronting physics, cosmology, and Buddhist philosophy. Thank you.

Our final moments together were filled with the exchange of gifts and expressions of gratitude. The Dalai Lama gave a long and beautiful white silk scarf to each of the presenters, together with an inscribed copy of his book The Power of Compassion. *Following a brief huddle, Anton Zeilinger invited His Holiness to visit him at his laboratories in Innsbruck, Austria. His schedule was immediately consulted, and three days were selected in June. And so the dialogue continues.*

The questions raised during our meeting have been addressed for millennia by scientists, philosophers, and contemplatives, and doubtless they will continue to occupy us for a long time to come. We were not attempting to answer questions of fact but rather were inquiring after the very nature of reality, the nature of knowledge, and the way in which we can create a better world. If, as the Dalai Lama says, the responsibility is in our own hands, we need to cultivate forms of knowing that reduce suffering for ourselves and for our planet.

The nearly miraculous accomplishments of modern physics and cosmology should not blind us to their implications. By pressing at the boundaries of natural science, we discover the mysteries of matter and energy, of space and time, of consciousness and world. The simplistic mechanistic models of the eighteenth century must be set aside, and we have found the newer models to be increasingly subtle and elusive. Our search for a robust and independent material reality on the other side of experience is frustrated again and again by the very facts of physics. Is there any reason to believe that relativistic quantum field theory or string theory will provide the perennially evasive substratum of reality? Or have we been thrown back on ourselves and on our experience in its fullest and most exact sense. Perhaps Buddhist philosophy is correct in seeking a middle way between the realism and the relativism. Others in the West have thought in similar ways.

I am reminded of Goethe's famous line "The facts themselves are the theory," which points out that a phenomenon—properly seen—is already theory, is already imbued with understanding. Can we learn to see our world and ourselves deeply, directly, and fully? Science will certainly help us in seeing accurately, but its finely polished and particular lens needs to be multiplied and enlarged to allow the full nature of our world to reveal itself. These dialogues were an attempt to do exactly this. Six scientist-philosophers joined the Dalai Lama for five days of unconstrained conversation. Every question was admissible; all approaches were sanctioned. For an inquiring soul, the exchange was thrilling. It is the way all inquiry should be.

NOTES

CHAPTER 2

1. Erwin Schroedinger, *What Is Life? And Other Scientific Essays* (Garden City, N.Y.: Doubleday, 1956), pp. 161–62.

2. See Chap. 3 of Paul Williams, *Mahayana Buddhism* (London: Routledge, 1989), for a treatment of Madhyamika. The Dalai Lama provides his own treatment of this subject in, for example, his commentary on the ninth chapter of Shantideva's *Guide to the Bodhisattva Way of Life,* which is published as *Transcendent Wisdom,* trans. and ed. by B. Alan Wallace (Ithaca, N.Y.: Snow Lion Press, 1988).

CHAPTER 3

1. Robert Hooke (1635–1703) performed early experiments in optics, in gravitation, and with the vacuum pump.

2. Until the mid-twentieth century, polarized light was analyzed by using certain crystals. Vikings used the polarization of the daytime sky to navigate. Recently, inexpensive sheet polarizers have become available. These are commonly used, for example, in certain types of sunglasses. Light from the blue sky is partially polarized.

3. Probabilities are always between zero and one (or we say 0 percent to 100 percent probability), and they are certainly never negative. This

would seem to require the matrix elements describing all operations to be positive. It turns out that in quantum mechanics this restriction can, indeed must, be relaxed to achieve sensible results.

4. Stated mathematically, the classical transformation equations between two reference frames in relative motion are

$$x = x' + vt' \text{ and } t = t'.$$

Note that the above equation for time does not include a reference to the spatial coordinate x. In Einstein's theory of special relativity, the transformation equations change to become

$$x = (x' + vt')/[1 - (v/c)^2]^{1/2} \text{ and } t = (t' + vx'/c^2)/[1 - (v/c)^2]^{1/2}.$$

Note that both equations have space and time coordinates on the right side. This is David's point.

CHAPTER 4

1. We need to recall the distinction in Buddhist philosophy between composite and noncomposite. The second hand sweeps through many different points as it passes around the dial. The interval of five seconds is "composed" of five one-second intervals, and so on. The spatial display of the clock face is composite. By contrast, noncomposite means that the entity under consideration has no parts. Empty space is totally free from obstructions of all types, and therefore no basis for division exists; that is, it is noncomposite. Apparently empty space may act like a medium through which objects or light may move; it must be considered composite.

2. The Vaibhasika tradition (also called Sarvastivada) was an early and important school in Hinayana Buddhism.

3. Kalachakra literally means "the wheel of time." It is generally recognized as the most complex and difficult of all Buddhist teachings and is especially studied in the Dalai Lama's own Namgyal monastery. See, for example, *The Wheel of Time: the Kalachakra in Context*, ed. Beth Simon (Ithaca, N.Y.: Snow Lion Publications, 1991).

4. See Bas van Fraassen, *The Scientific Image* (New York: Oxford University Press, 1980).

5. Andrei Linde, "The Self-Reproducing Inflationary Universe," *Scientific American*, November 1994.

CHAPTER 5

1. *The Life and Teaching of Geshe Rabten*, trans. and ed. B. Alan Wallace (London: George Allen & Unwin, 1980). See also F. Th. Tscherbatsky, *Buddhist Logic* (New York: Dover, 1962).

2. David is referring here to L. E. J. Brouwer, who, around 1912, founded the intuitionist school in mathematics.

CHAPTER 6

1. These two themes have been taken up in two other Mind and Life conferences. Daniel Goleman edited the conference book *Healing Emotions* (Boston: Shambala, 1997), which presents the November 1990 dialogue with the Dalai Lama. The March 2000 dialogue was published as *Destructive Emotions* (New York: Doubleday, Bantam, Dell, 2002), also edited by Goleman.

CHAPTER 7

1. Several months later, when Anton Zeilinger and I worked with the Dalai Lama in Innsbruck, Austria, Anton showed him this experiment in his laboratory.

2. Of course, in the ancient Greek period, Democritus, among others, advanced atomic theories, but these were considered speculative and heretical and were largely rejected in favor of Aristotle's doctrine of the four elements. Not until the seventeenth century did a scientific atomic theory arise.

ABOUT THE
MIND AND LIFE INSTITUTE

The Mind and Life dialogues between His Holiness the Dalai Lama and Western scientists were brought to life through a collaboration between R. Adam Engle, a North American lawyer and businessman, and Francisco J. Varela, a Chilean-born neuroscientist who was living and working in Paris. In 1984, Engle and Varela, who at the time did not know one another, each independently had the initiative to create a series of cross-cultural meetings in which His Holiness and scientists from the West would engage in extended discussions over a period of days.

After attending Harvard Law School, Engle worked at an entertainment law firm in Beverly Hills, California, and was for one year general counsel for GTE in Teheran. A restless spirit led to his first sabbatical year in Asia, where he became fascinated with the Tibetan monasteries he visited in the Himalayas. In 1974 he met Lama Thubten Yeshe, one of the first Tibetan Buddhists to teach in English, and spent four months at the Kopan Monastery in Kathmandu. When Engle returned to the United States, he settled near Santa Cruz, California, where Lama Yeshe and his teaching partner, Lama Zopa, had a retreat center.

It was through Lama Yeshe that Engle first became aware of the Dalai Lama's long-standing and keen interest in science and His Holiness's desire to both deepen his understanding of Western science and to share his understanding of Eastern contemplative science with Western scientists. Engle immediately felt that this was a project he would love to take on.

In the autumn of 1984, Engle, who had been joined on this adventure by his friend Michael Sautman, met with His Holiness's youngest brother, Tendzin Choegyal (Ngari Rinpoche), in Los Angeles and presented their plan to create a week-long cross-cultural scientific meeting, provided His Holiness would fully participate. Within days, Rinpoche reported that His Holiness would very much like to engage in discussions with scientists, and he authorized Engle and Sautman to organize a meeting. This began Tendzin Choegyal's continuing role as a key advisor to what is now the Mind and Life Institute.

Meanwhile, Francisco Varela, also a Buddhist practitioner since 1974, had met His Holiness at an international meeting in 1983 as a speaker at the Alpbach Symposia on Consciousness, where their communication was immediate. His Holiness was clearly happy for an opportunity for discussions with a brain scientist who had some understanding of Tibetan Buddhism, and Varela determined to look for ways to continue this scientific dialogue. In the spring of 1985, a close friend, Joan Halifax, then director at the Ojai Foundation, who had heard about Engle and Sautman's efforts, suggested that perhaps Engle, Sautman, and Varela could pool their complementary skills and work together. The four met at the Ojai Foundation in October 1985 and agreed to go forward. They decided to focus on the scientific disciplines dealing with mind and life as the most fruitful interface between science and the Buddhist tradition. This became the name of the first meeting and eventually of the Mind and Life Institute.

It took two more years of work among Engle, Sautman, Varela, and the private office of His Holiness before the first meeting was held in October 1987 in Dharamsala, India. During this time, Engle and Varela collaborated closely to find a useful structure for the meeting. Engle took on the job of general coordinator, primarily responsible for fundraising, relations with His Holiness and his office, and all other general aspects of the project; Varela, acting as scientific coordinator, took on primary responsibility for the scientific content, the invitations to scientists, and the editing of a volume covering the meeting.

This division of responsibility between general and scientific coordinators worked so well that it has been continued through all subsequent meetings. When the Mind and Life Institute was formally organized in 1990, Engle became its chairman and has been the general coordinator of all the Mind and Life meetings; although Varela has not been the scientific coordinator of all of them, until his death in 2001 he remained a guiding force and Engle's closest partner in the Mind and Life Institute and the series of meetings.

A word is in order here concerning the uniqueness of this series of conferences. The bridges that can mutually enrich modern life science, particularly the neurosciences, are notoriously difficult to engineer. Varela had a first taste of this when helping to establish a science program at Naropa Institute (now University), a liberal-arts institution created by Tibetan meditation master Chogyam Trungpa Rinpoche. In 1979 Naropa received a grant from the Sloan Foundation to organize "Comparative Approaches to Cognition: Western and Buddhist," probably the very first conference on that topic. Some twenty-five academics from prominent U.S. institutions gathered from various disciplines: mainstream philosophy, cognitive sciences (neurosciences, experimental psychology, linguistics, artificial intelligence), and of course Buddhist studies. The meeting provided a hard lesson to Varela on the care and finesse that organizing a cross-cultural dialogue requires.

Thus in 1987, profiting from the Naropa experience and wishing to avoid some of the pitfalls encountered in the past, Varela urged the adoption of several operating principles that have worked extremely well in making the Mind and Life series extraordinarily successful. Perhaps the most important was to decide that scientists would be chosen not solely by their reputations but also by their competence in their domain, as well as their open-mindedness. Some familiarity with Buddhism is helpful, but not essential, as long as a healthy respect for Eastern contemplative disciplines is present.

Next, the curriculum was adjusted as further conversations with the Dalai Lama clarified how much of the scientific background would need to be presented for His Holiness to participate fully in the dialogues. To ensure that the meetings would be fully participatory, they were structured with presentations by Western scientists in the morning session. In this way, His Holiness could be briefed on the fundamental knowledge of a particular field. The morning presentation was based on a broad, mainstream, nonpartisan, scientific point of view. The afternoon session was devoted solely to discussion, which naturally flowed from the morning presentation. During this discussion session, the morning presenter could state his or her personal preferences and judgments if they differed from the generally accepted viewpoints.

The issue of Tibetan-English language translation in a scientific meeting posed a significant challenge, as it was literally impossible to find a Tibetan native fluent in both English and science. This challenge was overcome by choosing two wonderful interpreters, one a Tibetan and one a Westerner with a scientific background, and placing them next to one another during the meeting. This allowed quick, on-the-spot clarification of terms, which is absolutely essential to move beyond the initial misunderstanding that results from two vastly different traditions. Thupten Jinpa, a Tibetan monk then studying for his *geshe* at Ganden Shartse monastery and now the holder of a Ph.D. in philosophy from Cambridge University; and B. Alan Wallace, a former monk in the Tibetan tradition with a degree in physics from Amherst College and a Ph.D. in religious studies from Stanford University, interpreted at Mind and Life I and have continued to interpret in further meetings. During Mind and Life V, while Wallace was unavailable, the Western interpreter was Jose Cabezon.

A final principle that has supported the success of the Mind and Life series is that the meetings have been entirely private: no press and no invited guests, beyond a very few. This stands in sharp contrast to meetings in the West, where the public image of the Dalai Lama makes a relaxed, spontaneous discussion virtually impossible. The Mind and Life Institute records the meetings on video- and audiotape for archival purposes and transcription, but the meetings have become a very protected environment.

The first Mind and Life dialogue was held in October 1987 in the Dalai Lama's private quarters in Dharamsala. Varela was the scientific coordinator and moderated the meeting, which introduced various broad themes from cognitive science, including scientific method, neurobiology, cognitive psychology, artificial intelligence, brain development, and evolution. In attendance, in addition to Varela and the Dalai Lama, were Jeremy Hayward (physics and philosophy of science), Robert Livingston (neuroscience and medicine), Eleonor Rosch (cognitive science), and Newcomb Greenleaf (computer science). The event was an enormously gratifying success in that both His Holiness and the other participants felt that there were some substantial advances in bridging the gap between cultures. Mind and Life I was transcribed, edited, and published as *Gentle Bridges: Conversations with the Dalai Lama on the Sciences of Mind*, ed. J. Hayward and F. J. Varela (Boston: Shambhala, 1992). This book has been translated into French, Spanish, German, Japanese, and Chinese.

Mind and Life II took place in October 1989 in Newport Beach, California, with Robert Livingston as the scientific coordinator and with the emphasis on neuroscience. Invited were Patricia S. Churchland (philosophy of science), J. Allan Hobson (sleep and dreams), Larry Squire (memory), Antonio Damasio (neuroscience), and Lewis Judd (mental health). It was during this meeting that Engle was awakened by a call at 3 A.M. with the news that the Dalai Lama had just been awarded the Nobel Prize for Peace and that the Norwegian ambassador would be coming at 8 A.M. to formally inform him of the award. After receiving the news, the Dalai Lama attended the meeting with the scientists as scheduled, taking time out only to hold a brief press conference about the prize. An account of this meeting is now available as *Consciousness at the Crossroads: Conversa-*

tions with the Dalai Lama on Brain Science and Buddhism, edited by Z. Houshmand, R. B. Livingston, and B. A. Wallace (Ithaca, N.Y.: Snow Lion Publications, 1999).

Mind and Life III returned to Dharamsala in 1990. Having organized and attended both Mind and Life I and II, Adam Engle and Tenzin Geyche Tethong, the secretary to the Dalai Lama, agreed that having the meetings in India produced a much better result than holding them in the West. Dan Goleman (psychology) served as the scientific coordinator for Mind and Life III, which focused on the theme of the relationship between emotions and health. Other participants included Daniel Brown (clinical psychology), Jon Kabat-Zinn (behavioral medicine), Clifford Saron (neuroscience), Lee Yearly (philosophy), Sharon Salzberg (Buddhism), and Francisco Varela (immunology and neuroscience). Daniel Goleman edited the volume covering Mind and Life III, entitled *Healing Emotions: Conversations with the Dalai Lama on Mindfulness, Emotions and Health* (Boston: Shambhala, 1997).

During Mind and Life III, a new extension of exploration emerged that was a natural complement to the dialogues but beyond the format of the conferences. Clifford Saron, Richard Davidson, Francisco Varela, Gregory Simpson, and Alan Wallace initiated a research project to investigate the effects of meditation on long-term meditators. The idea was to profit from the good will and trust that had been built with the Tibetan community in Dharamsala and the willingness of His Holiness for this kind of research. With seed money from the Hershey Family Foundation, the Mind and Life Institute was formed, which has been chaired by Engle since its inception. A progress report was submitted in 1994 to the Fetzer Institute, which funded the initial stages of the research project.

The fourth Mind and Life conference, "Sleeping, Dreaming, and Dying," occurred in October 1992, with Francisco Varela again acting as scientific coordinator. Invited participants were Charles Taylor (philosophy), Jerome Engel (medicine), Joan Halifax (anthropology; death and dying), Jayne Gackenbach (psychology of lucid dreaming), and Joyce McDougal (psychoanalysis). The account of this conference is now available as *Sleeping, Dreaming and Dying: An Exploration of Consciousness with the Dalai Lama*, ed. F. J. Varela (Boston: Wisdom Publications, 1997).

Mind and Life V, "Altruism, Ethics, and Compassion," was held again in Dharamsala in April 1995, and the scientific coordinator was Richard Davidson. The other participants included Nancy Eisenberg (child development), Robert Frank (altruism in economics), Anne Harrington (history of science), Elliott Sober (philosophy), and Ervin Staub (psychology and group behavior). The volume covering this meeting is entitled *Visions of*

Compassion: Western Scientists and Tibetan Buddhists Examine Human Nature, ed. R. J. Davidson and A. Harrington (New York: Oxford University Press, 2001).

Mind and Life VI opened a new area of exploration beyond the previous focus on life science. That meeting took place in October 1997, with Arthur Zajonc (physics) as the scientific coordinator The other participants, in addition to His Holiness, were David Finkelstein (physics), George Greenstein (astronomy), Piet Hut (astrophysics), Tu Weiming (philosophy), and Anton Zeilinger (quantum physics). The present volume covers this meeting.

The dialogue on quantum physics was continued with Mind and Life VII, held at Anton Zeilinger's laboratory at the Institut fur Experimentalphysic in Innsbruck, Austria, in June 1998. Present were His Holiness, Zeilinger, and Zajonc, as well as interpreters Jinpa and Wallace. That meeting was written up for a cover story in the January 1999 issue of *Geo* magazine of Germany.

The meeting described in this volume, Mind and Life VIII, was held in March 2000 in Dharamsala, with Daniel Goleman acting again as scientific coordinator and B. Alan Wallace acting as philosophical coordinator. The title of this meeting was "Destructive Emotions," and the other participants were the Venerable Matthieu Ricard (Buddhism), Richard Davidson (neuroscience and psychology), Francisco Varela (neuroscience), Paul Ekman (psychology), Mark Greenberg (psychology), Jeanne Tsai (psychology), Venerable Somchai Kusalacitto (Buddhism), and Owen Flanagan (philosophy). The volume covering this meeting is entitled *Destructive Emotions: How Can We Overcome Them?* ed. Daniel Goleman (New York: Bantam Doubleday Dell, 2002).

Mind and Life IX was held at the University of Wisconsin at Madison in cooperation with the HealthEmotions Research Institute and the Center for Research on Mind-Body Interactions. Participants were His Holiness, Richard Davidson, Antoine Lutz (sitting in for an ill Francisco Varela), Matthieu Ricard, Paul Ekman, and Michael Merzenich (neuroscience). This two-day meeting focused on how to most effectively use the technologies of fMRI and EEG/MEG in the research of meditation, perception, emotion, and the relations between human neural plasticity and meditation practices.

Mind and Life X took place in Dharamsala in October 2002 on "The Nature of Matter, the Nature of Life." The scientific coordinator and moderator was Arthur Zajonc (complexity), and the other participants, in addition to His Holiness, were Steven Chu (physics), Luigi Luisi (cellular biology and chemistry), Ursula Goodenough (evolutionary biology), Eric

Lander (genomic research), Michel Bitbol (philosophy), and Matthieu Ricard (Buddhist philosophy).

Mind and Life XI will be the first public meeting of this series. It will be held in Boston on September 13–14, 2003, and is entitled "Investigating the Mind: Exchanges between Buddhism and the Biobehavioral Sciences on how the Mind Works." In that meeting, twenty-two world-renowned scientists will join His Holiness in a two-day inquiry on how best to institute collaborative research between Buddhism and modern science in the areas of attention and cognitive control, emotion, and mental imagery. For more information on this meeting, see www.InvestigatingTheMind.org.

As an extension of the research begun in 1990, members of the Mind and Life Institute have again begun to research meditation in Western brain science laboratories with the full collaboration of meditation adepts. Using fMRI, EEG, and MEG, this research is being carried out at CREA in Paris, the University of Wisconsin in Madison, and Harvard University. Measures of emotional expression and autonomic psychophysiology are being gathered at the University of California at San Francisco and at Berkeley.

Paul Ekman of the University of California, San Francisco, a participant in the Destructive Emotions meeting in 2000, has developed a project entitled "Cultivating Emotional Balance." This is the first large-scale, multiphase Mind and Life research project designed to teach and evaluate the impact of meditation on the emotional lives of beginning meditators. The project has two primary research objectives: Design and test a curriculum to teach people to deal with destructive emotional episodes, drawing from Buddhist contemplative practices and Western psychological research, and evaluate the impact of the curriculum on the emotional lives and interactions of the participants. Alan Wallace, another participant in the 2000 meeting, has given advice on the development of the research project and is the meditation trainer. Ekman recruited Margaret Kemeny to lead the execution of the project, while he continues to provide guidance. The Fetzer Institute has provided initial support for the project, as well as a donation from His Holiness.

Mind and Life Institute
2805 Lafayette Drive
Boulder, CO 80305
www.mindandlife.org
www.InvestigatingTheMind.org
info@mindandlife.org

INDEX

big bang
  cause of, 182–183
  as centerless explosion, 173–175,
    178, 179–180
  inflationary model and, 99–100
  precursor to, 167
  space particles prior to, 88
  transformation theory and, 205
black holes, 69–70, 205
body and mind, 185–187
Bohm, David, 22, 27–28, 40
Bohr, Niels
  complementarity principle, 17
  Copenhagen interpretation, 29, 152
  minimalism of, 40
  objective picture of reality, 31
  observed phenomena, 25
  truths, simple and deep, 83
Boole, George, 53
Borromean rings, 25, 25f
brain, quantum events within, 186
breakthrough, 161–162
bridge metaphor, 200
Brouwer, L. E. J., 105
Brownian motion, 156
Buddha, omniscience of, 46
Buddhist thought
  analysis, modes of, 36, 110–111,
    112, 153
  astrophysics, parallels with, 99–100
  atomic theory, 132–133
  composite and noncomposite space,
    and absolutes, 76–77
  continuum, 188
  conventional reality (See conven-
    tional reality)
  deeds of an individual, 64
  delusion, sources of, 79, 81–82
  dependent origination, 32, 37,
    38–43, 105, 188–189
  elemental particles, 90–91,
    132–133
  emptiness (See emptiness)
  epistemology, 109, 111 (See also
    knowing and knowers)
  experience (See experience)
  Four Noble Truths, 97
  karmic causes, 48–49, 156–157,
    185
  karmic imprints, 183–184
  knowledge, positive orientation to-
    ward, 150–151 (See also know-
    ing and knowers)
  limitless universe and world sys-
    tems, 95–97, 99
  logic, 63, 101–102, 109, 202–203
  middle way (See Madhyamika)
  moral choices and omniscience, 46
  Nagarjuna (See Nagarjuna)
  negation process, 38, 87, 160
  observation (See observation and
    the observer)
  origins of things, position on,
    182–183
  oscillating universe hypothesis,
    93–96
  phenomenology (See phenomenol-
    ogy, Buddhist)
  science and, 6–7
  space, nature and classifications of,
    87–88
  substantial vs. imputed entities,
    85–86
  time, nature and classifications of,
    85–86, 88–89
  unanswered questions, 220–221

calcite, 56
Calculator, Divine, 44–45
causality
  classifications of, 185–186
  dependent origination and, 37,
    188–189
  determinism, 43–46, 184, 187–188
  free will, 183–185
  hidden, 26, 27, 47
  karmic causes, 48–49
  as level of analysis, 127
  microscopic causality, failure of,
    57–58
  mind, body, and karma, 185–187
  randomness and, 20, 26–28,
    33–36, 185–186
  simultaneous, 35, 189
  speed of light as constraint on,
    187–188
centerless explosion, 174–175, 178,
  179–180
central nervous system, and sense
  organs, 144

star formation, 96
states
    idea of, 52–54
    intrinsic reality and, 92
    measurements assumed in, 58
    reality and, 64
    superposition, 103
statistical character of quantum
    mechanics, 186
strangeness, 47, 49
structure of atoms, 134
structures of galaxies, 177
students, qualified, 216
subjective randomness, 19, 33, 43
subjectivity, and science, 208,
    209–210
subject-object duality, 211–212
substantial entities, 85–86
suffering, 97, 220
superposition
    double-slit experiment and, 116,
        117f
    excluded middle and, 102–104
    as "fruitful ambiguity," 123
    knowledge of, 138–139
    understanding of, 151–152
syllogistic reasoning, 113

tactile objects, 133
Taoism, 34, 90–91, 217
telescopes, 165
temperature and motion, 135–136
theory, relationship with experience,
    126, 134, 154–155, 161–162,
    203–204
thermodynamics, 78
Thomson, William (Lord Kelvin), 4
thought experiments, 22
Tibetan Buddhism. *See* Buddhist
    thought
Tibetan Children's Village, 3
Tillich, Paul, 149
time
    chronons as least unit of, 70–71
    defining, 77
    infinity of universe and, 180
    nature and classifications of, Bud-
        dhist, 85–86, 88–89
    relativity of simultaneity, 79–84,
        80f

thermodynamic basis for time's
    arrow, 78
    See also space-time
train examples, 66–67, 80–82, 80f,
    205
transformation matrix, 52–53
transformation theory, 204–207, 209
transmutation, 205
trekchö (breakthrough), 161–162
truth
    absolutes and, 105
    action and, 194
    as goal, 7
    moral vs. scientific realm,
        149–150
    simple and deep, 83–84
    See also knowing and knowers;
        logic; reality
Tsong-khapa, 92

ultimate analysis, 112, 152–153
ultimate reality, 156, 159–162
unbounded universe, 168–171
uncertainty principle, 45, 69, 142
unified field theory, 191
universality, 18–19, 110, 216
universals and indistinguishability,
    106–107
universe. *See* cosmology

Vaibhasika school, 87–88, 224n2
Vajrayana school, 96, 161
validation of cognition
    within conventional reality,
        152–153, 159–160
    illusory knowledge, 211
    in Madhyamika vs. realist schools,
        107–109
    meditative training and, 212
    theory and experience, 161
    validation of objective world and,
        104–105
value judgments, 109
Vasubhandu, 97
velocity, relative, 172–174
verbal designation, 37
vertical polarization, 21
Vikings, 55, 223n2
void, 90
voids, 177